A12900 143610

D1154759

TK
5102
.2
.L46
1979

LEINWOLL
From spark to satellite.

WITHDRAWN

Illinois Central College
Learning Resources Center

FROM SPARK TO SATELLITE

FROM SPARK TO SATELLITE

A HISTORY OF RADIO COMMUNICATION

Stanley Leinwoll

Edited by **FRED SHUNAMAN**

42926

CHARLES SCRIBNER'S SONS / NEW YORK

I.C.C. LIBRARY

Copyright © 1979 Stanley Leinwoll

Library of Congress Cataloging in Publication Data
Leinwoll, Stanley.
From spark to satellite.
Bibliography: p. 233.
Includes index.
1. Telecommunication—History. I. Title.
TK5102.2.L46 384'.09 78-24172
ISBN 0-684-16048-X

This book published simultaneously in the
United States of America and in Canada —
Copyright under the Berne Convention

All rights reserved. No part of this book
may be reproduced in any form without the
permission of Charles Scribner's Sons

TK 5102 .2 .L46

1 3 5 7 9 11 13 15 17 19 V/C 20 18 16 14 12 10 8 6 4 2

Printed in the United States of America

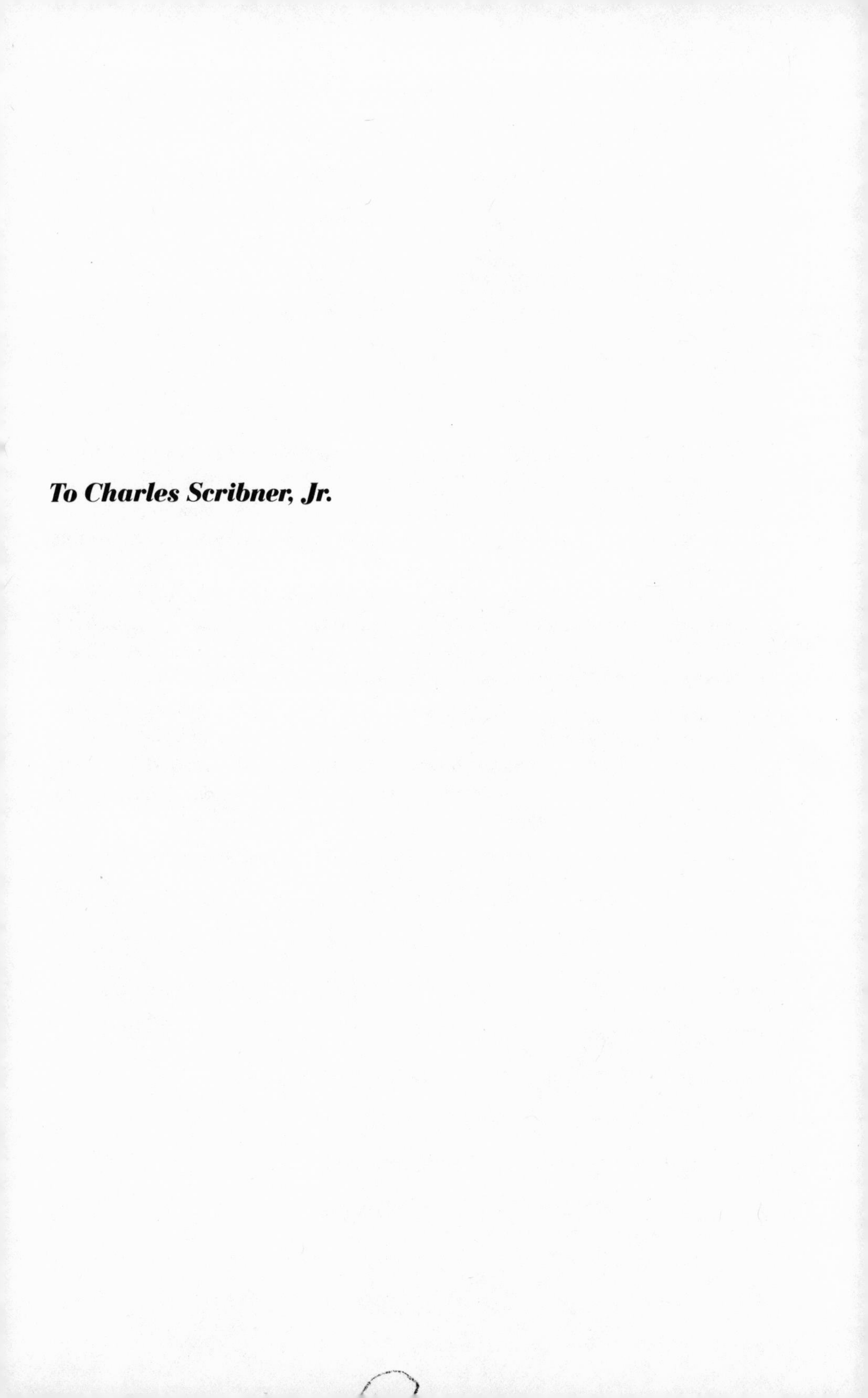

To Charles Scribner, Jr.

Acknowledgments

I am grateful to W. J. Baker, Technical Editor of the Marconi Company, Ltd., for providing valuable information about the early days of wireless, and for commenting on certain sections of the manuscript. Richard Baldwin of the American Radio Relay League supplied considerable material dealing with amateur radio. Valuable contributions to the book were made by officials of NASA, COMSAT, RCA Corporation, General Electric Company, Bell Telephone Laboratories, American Telephone and Telegraph Company, and Westinghouse Broadcasting Company. Art Trauffer of the De Forest Museum in Council Bluffs, Iowa, was very helpful supplying photos and information about Lee de Forest. Many others assisted in the preparation of the manuscript, and I am grateful to all. A special word of thanks goes to my wife, Miriam, who read through the manuscript, and made many valuable suggestions which, I am certain, strengthened the book.

Contents

Introduction

The ability to communicate thoughts and ideas over time and space is one of the features that distinguishes man from other animals. If humans had been unable to exchange information they would still be roaming the forests, plains, and jungles, foraging for food, struggling to survive in their hostile environment. From the time man first stood on his hind legs and began to walk erect, he has devoted much of his time to improving his means of communicating. Once he had learned to make intelligible sounds he turned to the task of representing his ideas with drawings, then with more abstract symbols. Next came the task of transmitting intelligence over distances. Short-range communication was limited to his line of sight, and he used the drum, smoke, and fire. Such short-distance communication was essentially instantaneous. The speed at which man could communicate over distances beyond the horizon was generally limited to the means of transportation at his disposal: animals, sailing vessels, and foot. Later he used carrier pigeons.

The limitations on man's ability to exchange information made his scientific and technological progress painfully slow. Even after the invention of the printing press the dissemination of printed material was limited.

Then—after centuries of tediously slow development—man suddenly broke through with the invention of the telegraph and telephone. The door to rapid, worldwide communication between peoples had been opened. But although he could now begin to communicate rapidly over great distances, he needed wires and cables, which were expensive and time-consuming to lay and which still limited the areas of communication.

It was the genius of Guglielmo Marconi that gave man the freedom to communicate instantaneously between any two locations on earth, whether they were connected by wires or not. Through the miracle of radio waves, man's thoughts could literally soar into space, moving invisibly with the speed of light, carrying the sum total of all man's knowledge and accomplishments. Wireless, or radio, communication is one of the major developments in the history of science. Communication before the invention of wireless can be compared to transportation before man could fly.

Wireless communication is defined as the transmission and reception, over great distances and without intervening wires, of electromagnetic waves that are translated into visual or aural messages.

The early development of wireless technology is clouded in controversy. Marconi was the first to realize that control of the medium lay in commercial application and the protection of patent rights. In the struggle to control the priceless inventions of wireless, bitter controversies arose among the early scientific giants of the medium: Lee de Forest, John Ambrose Fleming, Reginald Fessenden, Edwin H. Armstrong, and Marconi were all involved in titanic struggles, some of which lasted for decades. The early inventors also found the going difficult because the concept of wireless communication was not easily grasped, and money for research and development was difficult to obtain; some eminent scientists were quick to assert that such communication was impossible.

The development of wireless communication was uneven, with the first twenty years showing painstakingly slow progress. This was followed by a rapid expansion, then another period of slow development, followed by a period of unprecedented progress.

Today we are immersed in radio waves. Without our perceiving them, thousands of signals inundate us: broadcast and commercial radio signals, television, shortwaves, amateur transmissions, signals that would be unintelligible to us even if our senses could detect them because they are coded—data, facsimile, photos. With invisible, ubiquitous radiation, radars probe the skies for missiles and aircraft. Satellites hover thousands of miles above the earth, relaying phone conversations, radio programs, television signals; computers speak to other computers thousands of miles away. Other communication systems are en route through space to the planets to return information about the universe in which we live.

Radio waves come to us from the planets, the sun, other stars, and from mysterious bodies in the heavens. These signals cannot be seen, heard, or felt without sophisticated instruments to detect them and to translate the complex messages they send to us. These messages tell of the origin of the universe, of the birth and death of stars and galaxies, of cataclysmic explosions and unimaginably vast turbulent clouds, bubbling and boiling at the edge of the infinite.

We have moved, in the briefest interval, from using several hundred kiloHertz to virtually the entire electromagnetic spectrum. Even visible light in the multimillion megaHertz region is within reach as a practical, usable part of the spectrum for communications.

The miracle of radio becomes even more remarkable when we consider that this strange and fascinating form of electromagnetic radiation was first postulated barely one hundred years ago by James Clerk Maxwell. It was Heinrich Hertz, little more than a decade later, who first proved that these mysterious waves really existed, in one of the greatest experiments in the history of science.

Modern communication consists of complex, integrated systems composed of many devices and technologies. These have shaped the pattern of our social, political, cultural, and economic development, accelerating it as in no other period of human history. By its nature, communication has become the greatest unifying force ever to work for man. It has brought men closer, giving them a better understanding of the world and of how others live and function. It has revolutionized our lives.

FROM SPARK TO SATELLITE

1

The Father of Wireless

On December 12, 1901, on a hill overlooking St. John's, Newfoundland, an event took place that was destined to have a profound effect on the social, cultural, political, and economic life of all peoples on earth from that day onward. At 12:30 P.M. on that cold and blustery day, a handsome young man of twenty-seven was working at a table on which an unusual collection of electrical equipment was assembled. The building that housed the apparatus barely sheltered him from the harsh winds that blew outside.

Holding a telephone receiver tightly to his ear, the young man listened intently. Suddenly his expression brightened. Beckoning to his assistant, who was waiting nearby, he handed him the receiver. "Can you hear anything, Mr. Kemp?" he asked.

Kemp pressed the telephone to his ear. Listening for several seconds, he smiled and nodded affirmatively, handing back the receiver. Both had heard three faint clicks in the receiver—Morse code for the letter S. The energy that produced the signals had traveled more than 2,000 miles from Poldhu, near Land's End, in Cornwall, England. The two men, Guglielmo Marconi and George Kemp, heard the signals again at 1:10 and 2:20 P.M. the same day, and at 1:28 P.M. the following day, December 13.

The announcement—given to the press on December 16—startled the world. Electrical signals had been sent across the Atlantic without wires! The achievement—one of the significant steps forward in human history—climaxed seven years of work by the Italian scientist. The world would never be the same again.

Guglielmo Marconi was born on April 25, 1874, in Bologna, northern Italy. His father, Giuseppe, was a well-to-do businessman; his mother, the former Annie Jameson, was Irish. She had been born in Dublin, the daughter of Andrew Jameson, who operated one of Ireland's largest whiskey distilleries. Annie had come to Italy to study *bel canto*. There she met, and later married, Giuseppe.

As a child, Guglielmo had few friends. At the age of five he went to England for two years. His first elementary school education was at a private school in Bedford. For the next several years most of his education was provided by tutors and by his mother, who taught him in English. He went to school in Florence at age twelve, and at thirteen attended the Leghorn Technical Institute. Signora Marconi provided Guglielmo's religious training and also taught him music. He became an accomplished pianist, but his first love was science.

He read extensively in the family library, and in his teens attended some of Professor Augusto Righi's lectures. Righi was Italy's leading authority on electromagnetic radiation. The lectures stimulated Marconi's interest in electrical phenomena, and by the time he was twenty he had read extensively on the subject.

The turning point in Marconi's life came at the age of twenty, when he read, while on vacation in the Italian Alps, a paper by Righi on the experiments of Heinrich Hertz, who had just died. Using a battery, an induction coil, a switch, and a pair of metal plates with a spark gap between them, Hertz had produced an electrical discharge, which was detected several feet away by a circlet of wire with a small gap in it. When the discharge across the main "spark gap" occurred, tiny sparks could be seen across the gap in the circlet. Energy had been transferred through space!

His imagination fired by the article, young Marconi curtailed his vacation and returned to the family's country estate, the Villa Griffone, near Bologna. Signora Marconi had given her son a room on the third floor of the house to use as a workshop and laboratory. Here Guglielmo conducted his first experiments in communication with electromagnetic waves. His early efforts consisted of modifying the Hertzian apparatus in an attempt to produce bigger sparks at greater distances in his receiving circlet. He had seen the possibilities: Electric waves might be used to transmit and receive messages over great distances without wires!

It was not long before Marconi was able to transmit the full length of the room. It then became clear to him that further development would have to lie in two directions: He would have to increase the distance the sparks could be transmitted, and he would have to make those sparks transmit information in some manner. Young Marconi realized that this would take capital, and went to his father for it. The elder Marconi was totally against his son's activities at first, but soon saw the commercial possibilities of his son's "wireless" experiments, and gave the boy small sums of money to continue his work.

Other researchers of the day in England, Germany, the United States, and Russia were conducting experiments with Hertzian apparatus, but none of them appear to have been very successful at the time Marconi entered the field. He started with equipment used by the others. "I reproduced with rather rudimentary means an oscillator similar to that used by Righi," he says. "[I] likewise reproduced a resonator using as a detector of the electrical waves a tube of glass with pulverized metal based on much that was already published by Hughes, Calzecchi-Onesti, Branly and Lodge."

The "tube of glass with pulverized metal" had been developed by a French physicist, Edouard Branly, who at the time of his original experiments did not realize that Hertzian waves, generated by the spark discharge, were responsible for the cohering effect he observed. It was widely believed at the time that the light from the spark was producing the effect. It was Sir Oliver Lodge who, in a lecture delivered in June 1894, was the first to demonstrate clearly that the Branly tube could detect Hertzian waves. Lodge christened the device a "coherer."

Marconi then developed a crude decoherer—a "clapper" that would "touch the tube every time I sent a train of electric waves." With the clapper he was able to control the length and spacing of his signals. "Precisely at that time," he said, "I thought for the first time of transmitting telegraphic signals and substituting a Morse machine for the voltmeter" (used for a detector). "The very weak current available with my materials was insufficient to make a Morse machine function. I at once thought of reinforcing the current with a relay. And this I did later."

Marconi's equipment worked, but only over distances comparable to those attained by other researchers at the time—a matter of yards at the most. Then a lucky event occurred. He had been using a pair of large metal sheets (presumably originally used as reflectors) attached to the terminals of his transmitter, apparently to increase the effect of the plates attached to his spark balls. They made his next step forward possible. In his own words: "I found out how to obtain waves at distances of hundreds of meters. By chance I held up one of the metal slabs at a considerable height from ground and set the other one on the earth. That was when I first saw a great new way open before me."

Marconi soon found that the distance he could send a signal varied approximately in proportion to the length of the two vertical wires, as well as the height of the plates above ground. He also found that his coherer could be improved. The Branly tube was too erratic to receive Morse signals reliably. So he experimented with more than three hundred combinations of filings and metals to evolve a satisfactory coherer.

Each impulse reaching Marconi's receiving equipment produced the same result: the particles cohered, current flowed, the tapper struck the tube containing the particles, which decohered and were ready to receive another impulse. If the spark transmitter was "keyed" to send longer and shorter signals, the decoherer continued to tap during the length of each signal, making it possible to read a Morse message from it. Using the device, Marconi was able to transmit dots and dashes over a distance of about 1 mile.

He soon discovered that if the receiving equipment was placed on the far side of a hill, signals could still be received. This indicated that the waves were either going through the hill, or traveling over it. At this point Marconi's family and some close associates became very much interested in the dramatic results he was getting. Marconi did not, however, publicize his findings because he had no patent protection.

In 1895, at the age of twenty-one, Marconi offered his invention to the Italian Ministry of Posts and Telegraphs, which turned it down because there seemed to be no particular use for it. He was told his equipment would be more useful to a maritime nation because it seemed to lend itself more to communication between ships, or between ship and shore. At that time England was the world's most powerful maritime nation. Since his mother had friends and influential acquaintances in that country, it seemed a logical place to take his wireless.

Marconi arrived in London in February of 1896 with two trunks of wireless equipment. Customs officials were suspicious of the young Italian immigrant. Fearful that he might be an anarchist and that his mysterious apparatus might be an infernal machine, they proceeded to dismantle the equipment completely. They were not able to put it back together again, and damaged some of the parts in the process. Marconi had to make hasty repairs before he could demonstrate his new technique.

The first man to see the Marconi wireless operate in England was his cousin, Henry Jameson Davis, an influential businessman in his own right. Plans were made to patent the invention. On June 2, 1896, Marconi applied for a patent for his wireless telegraph invention. By this time he had already contacted several prominent engineers in Great Britain. Among them was William Henry Preece, Chief Engineer of the British Post Office. Preece had conducted his own researches in telegraphy and had tried to approach wireless by the use of magnetic induction techniques, with limited success. Acting as a watchdog for the Post Office, he was keeping an eye on whatever might turn out to be a rival system to the Post Office's wire-conductor telegraph. He offered to assist the young man in any way he could. Marconi was interested in the Post Office because, having a monopoly on all means of communication in Britain, it was potentially a valuable customer. The two therefore formed an association of mutual convenience.

Marconi demonstrated his equipment to officials of the Post Office and the War Offfice in July and August 1896, transmitting signals to distances of several hundred yards. This brought a request for further demonstrations, and the equipment was moved to Salisbury Plain, where successful communication was established over a distance of a mile and three-quarters. Subsequent tests extended that distance to 4 miles, and a

Fig. 1. Marconi and early wireless apparatus, 1896. *Courtesy of the Marconi Company.*

test across Bristol Channel to 8 miles. But Marconi believed that—though the distances covered were remarkable—the limit had by no means been reached.

Marconi's experiments attracted the immediate attention of the press, and Marconi found himself heralded as "the inventor of wireless." Actually many eminent scientists had made important contributions in the field of Hertzian waves, their behavior and their possible use in communication. Some of them were outraged at the fact that Marconi was receiving all the credit for the invention, with no mention of the fact that he was using their equipment. In particular, the eminent electrical engineer Sylvanus Thompson attacked him in a letter to the *Saturday Review* of London, stating that Marconi was using, with the exception of a few details, "the system of Lodge," who was "the original inventor of wireless telegraphy." Marconi replied in the same paper that Thompson's description of the system as Lodge's was somewhat unfair, "in view of the previous work of the late Professor Hughes, of Branly, Popov and others," and went on to say that the few details in which his system differed "made the whole difference between workableness and unworkableness in a wireless telegraph apparatus of the type in question."

Marconi continued to experiment and to demonstrate his equipment. In mid-1897 he returned to Italy for a demonstration of his apparatus to the Royal Italian Navy. The test covered a distance of more than 10 miles, and as a result the Italian navy decided to adopt the Marconi equipment to communicate with its ships at sea.

While Marconi was in Italy, his cousin Henry Jameson Davis formed the Wireless Telegraph and Signal Company, Ltd. It had become clear to Marconi, after the first series of test demonstrations to the British Post Office had been concluded, that the Telegraph Acts of 1868 and 1869 made it impossible for him to organize an internal message-carrying service for gain in England; thereafter he would turn his attention to maritime services because the Post Office had no monopoly in those areas.

Marconi used his wireless in 1898 to report the Kingstown Regatta yachting races for the Dublin *Express*. He followed the racing yachts in a tug that had been equipped with wireless apparatus, and radioed to shore a running commentary of race positions and developments. This was probably the first time wireless was used for journalistic purposes, and—since the paper paid for the service—the first commercial radio transmission.

Queen Victoria was so taken with the Kingstown Regatta reporting that she requested that wireless communication be established between her residence at Osborn House, on the Isle of Wight, and the Prince of Wales, who was recovering from an injury aboard the royal yacht *Osborne*, several miles away. More than one hundred messages were exchanged between the Queen and the Prince, and the exchanges received wide newspaper coverage.

In 1898 Marconi's company had more than $500,000 in assets, and with Marconi as chief engineer, enjoyed a virtual monopoly of wireless manufacture in England. Although the British navy and War Office provided some business for the company, it did not satisfy Marconi. Aware of a prediction by Sir William Preece that the new system of wireless communication would reach hitherto inaccessible places, and that for shipping and lighthouse purposes it would be a great and valuable asset to humanity, he concentrated his attention on installing wireless equipment in lighthouses. He knew that attempts to lay cables between shore points and lighthouses had proven fruitless because of the battering from heavy seas.

One of the first projects the new Wireless Telegraph and Signal Company undertook was a series of tests between a lighthouse on Rathlin Island, off Ireland, close to the shipping lanes, and a telegraph office at Ballycastle, 7 miles away on the mainland. Tests were also conducted between a ship at sea and a lighthouse about 12 miles away. Wireless equipment was installed on the East Goodwin lightship and several dramatic sea rescues followed; in March and April 1899 the lives of two ships' crews were saved when the ships went aground near the lightship.

In 1899 the French government requested that Marconi conduct tests to determine whether communication between England and the European continent was feasible. The tests, carried out over a 30-mile distance, were a complete success, and were given wide publicity by the many reporters from both countries who witnessed them. (It is said that Marconi's first message to the Continent congratulated Edouard Branly on the success of his instrument, the coherer.) At last wireless was beginning to gain international attention.

In the same year Marconi came to the United States to conduct a series of tests for the War and Navy departments. The American military adopted Marconi's system for use by the army and navy.

While in the United States, Marconi gained widespread publicity by reporting the results of the America's Cup yacht races off Sandy Hook. As *Shamrock* and *Columbia*, contenders for the historic cup, sailed the course, Marconi reported the results from a tug, describing the events in detail as they unfolded. It was the first time a sporting event in the United States

Fig. 2. Marconi at Signal Hill, Newfoundland, with the instruments with which he received the letter S, three dots in the Morse code, sent from Poldhu, Cornwall, on December 12, 1901. *Courtesy of the Marconi Company.*

had been covered by wireless. *Shamrock*, entry of Sir Thomas Lipton, British tea magnate and grand old man of yachting, lost to the American entry.

Although Marconi received important attention in the press during his stay, and the U.S. government purchased his equipment, the most significant event of his visit was the formation, in November 1899, of the Marconi Wireless Telegraph Company of America. (Two decades later it was to become the Radio Corporation of America.)

At twenty-five, Guglielmo Marconi had gained what most men fail to achieve in a lifetime—international recognition and respect, moderate wealth, and a place in history. But his greatest moments still lay ahead. By 1900 he was experiencing serious competition from foreign sources. Most active were the Germans—Ferdinand Braun, of the major firm Siemens and Halske, and Adolph Slaby and George von Arco, of AEG, the German General Electric Company. (Later the wireless interests of the two firms were pooled in the Telefunken Company.) The head start Marconi had gained was being held, but others were close behind and he needed some innovation that would give him a significant lead.

One of the biggest problems of the time, as competition increased, was co-station interference. Because nobody tried to control the frequency at which the wireless equipment worked, two stations operating close together often drowned each other out. In 1900 there was no way of separating a wanted from an unwanted signal, and since there was no regulation, conditions were often chaotic. Receiving stations often received only a hodgepodge of incoming signals from two or more transmitters. "Tuning" was unknown; transmitting and receiving stations resembled those of Figures 3 and 4.

The solution to this problem of interference was found in the use of tuned, or resonant, circuits. The principle of resonance, called *syntony* by

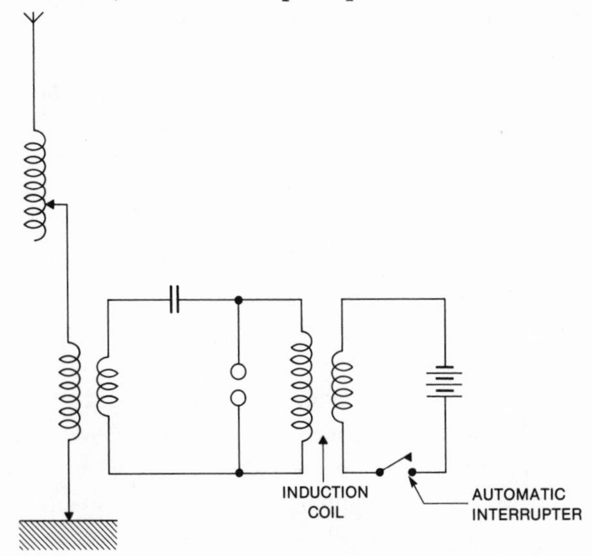

Fig. 3. Marconi transmitter, about 1900. *From* A History of the Marconi Company, *W. J. Baker*.

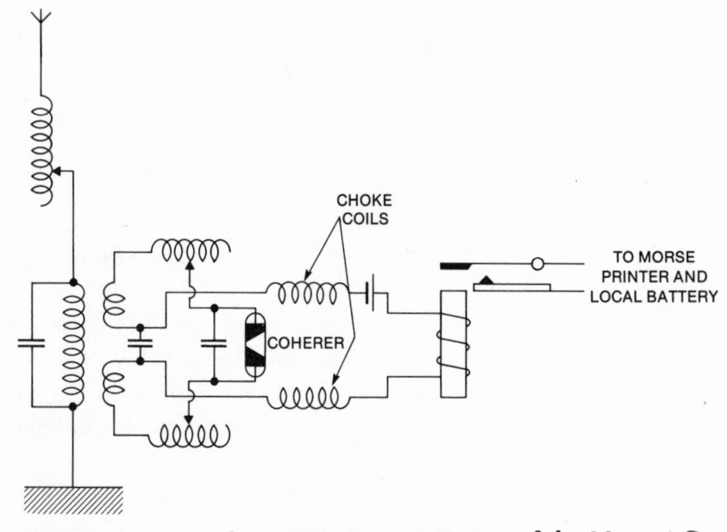

Fig. 4. Marconi receiver, about 1901. *From* A History of the Marconi Company, *W. J. Baker*.

Sir Oliver Lodge (who demonstrated it in 1897), made use of virtually identical antennas, inductances and capacitances in both transmitter and receiver. Braun had patented a similar device in 1899, and Tesla and John Stone Stone had worked out similar ideas in the United States. But the systems that Lodge and Braun had patented had one serious drawback—very little energy was radiated into space. Two simple yet ingenious innovations by Marconi solved the problem. He coupled the antenna *inductively* to the transmitter (used two coils placed in close relation to each other instead of a single coil or none at all, as in some earlier transmitters). He also made the antenna inductance, and the capacitance in the transmitter circuit, variable. The changes enabled him to *tune* his transmitting and antenna circuits to resonance with each other. No longer did his wireless equipment radiate a broad band of frequencies, and the oscillations radiated considerable energy into space.

Marconi then matched his receiving circuits to those at the transmitter, tuning to the frequency being transmitted. By using syntonic circuits with variable inductance and capacitances, stations could operate in the same vicinity, and by varying the circuit components, could transmit and receive with greatly reduced interference. As soon as he was certain that syntony was the answer, Marconi applied for an all-inclusive patent on his system. On April 26, 1900, he received one of the most important patents ever granted, the famous No. 7777.

Coincident with the work on syntonic, or tuned, circuits, Marconi had become aware of an apparent paradox in his wireless experiments. He knew that, according to the well-understood laws of electromagnetic wave propagation, wireless telegraphy distances should never greatly exceed optical distances. Radio waves—like light waves—travel in straight lines. Therefore, because of the earth's curvature, they were expected to leave the surface at a tangent and disappear out in space. Diffraction and refraction would increase the range to a little beyond the horizon, but no significant extension could theoretically be expected, according to the best scientific knowledge of the time.

Scientists of the day were therefore unanimous in declaring that wireless telegraphy would be limited to just-beyond-the-optical-horizon distances. But Marconi found that in practice he was obtaining ranges at least twice those mathematical calculation would indicate. He did not know why, he only knew it was so.

Encouraged by the 60–100-mile ranges he was already getting, Marconi decided to gamble by seeing whether the signals could bridge the Atlantic. The audacity of this scheme can be imagined by remembering that all wireless equipment at that time was small and battery-powered. A transatlantic project would demand a huge power plant and an antenna system of a kind never before visualized. Marconi proposed to erect two such stations—one each side of the Atlantic—establish two-way communication, and thereby break the monopoly of the cable companies.

He put his proposal to his board of directors. Not unnaturally, they were far from enthusiastic. Tremendous expenditure would be involved and, according to the scientific texts, the idea was unworkable and could never succeed. At length, however, and with considerable misgivings, the board agreed.

When the news was released many scientists scoffed. The Earth was round; Hertzian waves traveled in straight lines; the signals would be lost in outer space long before they reached their destination. There was no way, they said, that the experiment could succeed. But Marconi was stubborn and would not be dissuaded.

To assist in the project, Marconi engaged an eminent scientist and engineer, J. A. Fleming, professor of electrical engineering at London University. Fleming, who would later invent the vacuum-tube diode (the famous Fleming valve), was an expert in high-power alternating-current generators, and was an authority on the work of Maxwell and Hertz as well. He had duplicated Hertz's experiments and had followed Marconi's work closely.

In mid-1900 a site at Poldhu, just east of Land's End, Cornwall, was chosen for the transmitter site. Construction was begun in October 1900, the work being carried out secretly. The Poldhu station was massive. Instead of battery power supplies, a 32-horsepower generator was installed. It drove a 25-kilowatt alternator, which had an output of 2,000 volts. This voltage was stepped up to 20,000 by a transformer.

The antenna system at Poldhu consisted of twenty wooden masts, each 200 feet high, erected in a circle. The masts were stayed by horizontal triatics (cables between the tops of each mast, from which the antennas were suspended) and the guy wires anchoring the masts were broken up with insulators to reduce absorption of radiated power. From an engineering standpoint, the horizontal support arrangement was a bit doubtful. If one mast were to fall, it would probably bring all the others down with it. In view of the low losses of the triatic arrangement (the triatic cables were almost at right angles to the antenna conductors), the designers decided to gamble. They would need every bit of power they could get.

By March 1901 the Poldhu station was nearly ready, and Marconi sailed with an assistant for the United States. An oceanfront site on Cape Cod, at South Wellfleet, Massachusetts, was chosen for the American station. Leaving the construction of the receiving station and antenna to his assistant, R. N. Vyvyan, he returned to England.

Preliminary tests conducted in the summer of 1901 from Poldhu to Crookhaven, on the west coast of Ireland, were successful. The distance was 225 miles. This was well beyond the 186-mile record established from Poldhu several weeks before, and was further evidence that the Hertzian waves were following the curvature of the Earth. Since no one had any idea of "reflecting layers" in the upper atmosphere at that time, Marconi and his associates supposed the signals were traveling along the Earth's surface.

During these preliminary tests, construction of the Cape Cod station continued, and the finishing touches were being put to the Poldhu installa-

Fig. 5. Poldhu wireless station in 1901 with fan-shaped aerial. *Courtesy of the Marconi Company.*

tion. Then disaster struck. On September 17, in one of the worst gales ever to strike the area, a lug was torn from the top cap of one of the masts, releasing the end of the triatic attached to the mast. All one hundred masts fell immediately, in a mountain of twisted debris.

Marconi immediately ordered the wreckage to be cleared and an alternate temporary antenna system to be erected. Within the month the site had been cleared and the second array up and under test. It consisted of sixty copper wires arranged in a fan shape and mounted between two 150-foot masts. (Marconi used 7-22 bare copper wire, presumably because it was available, and for decades thereafter 7-22 bare copper was sold as the "standard" antenna wire.)

While the tests on the temporary antenna were going on, plans were made to substitute a more powerful permanent one. But the temporary one worked out so well that Marconi decided not to wait, but to use it for the transatlantic tests. But now a second disaster, identical to the first, befell the Cape Cod antenna. It collapsed in November during a northeastern storm.

Deciding that he could not wait to reconstruct the stateside system, Marconi set sail for the nearest landfall in the Americas—Newfoundland. With him were two assistants, George S. Kemp and P. W. Paget. They carried with them an assortment of wireless equipment, including different receivers and coherers, antenna accessories—including large canvas kites, balloons, and varying lengths and thicknesses of wire.

Landing at St. John's on December 6, 1901, Marconi met with Sir Cavendish Boyle, Governor of Newfoundland, and Sir Robert Bond, Prime Minister. Both of them promised him all possible assistance. They made available to the party an abandoned barracks hospital, which lay on a hill some 600 feet above St. John's harbor, facing Poldhu. The location, now called Signal Hill, was near the point where John Cabot, first-known white man to set foot in Newfoundland, first landed.

By December 9 Marconi and his assistants had finished assembling their equipment in a ground-floor room of the hospital. Marconi sent a cable to Poldhu, instructing the technicians there to start transmitting test signals on December 11. They were to send the letter S—three dots in the Morse Code—continuously between 11 A.M. and 3 P.M. Newfoundland time. The letter S was chosen for several reasons. The switching arrangements at Poldhu would not withstand long periods of operation, especially if dashes were sent. Furthermore an automatic device could be used to send S's. Finally, Marconi felt that three dots would probably be heard most easily if atmospheric noise was heavy.

An antenna carried by a large kite was sent up on December 10, in preparation for the tests the following day. Everything went smoothly. Then, on December 11, the weather deteriorated. By midmorning a full gale was blowing. Attempts to send an antenna aloft failed and a kite and balloon were lost in the severe winds that lashed the hillside. No signals were heard that day.

The next day—December 12, 1901—the weather continued harsh, with a full gale raging. In spite of the weather a kite-carried antenna was sent to an altitude of 400 feet, and Marconi began his listening vigil. But the winds made the motion of the kite highly erratic, dipping and soaring like a terrified bird. These movements changed the angle the antenna made with the Earth. As a result the antenna characteristics were in a constant state of flux. Marconi heard nothing.

He had substituted a telephone receiver for the Morse inker he had been using. The inker would have given him a printed record of the experiment, but was not nearly as sensitive to signals as the human ear. He also replaced his syntonic receiver with an older, untuned model. Different coherers were used; one of these was the so-called Italian navy coherer de-

Fig. 6. Raising the kite aerial at Signal Hill, Newfoundland, in 1901. Marconi is at the extreme left. *Courtesy of the Marconi Company.*

veloped by Marconi's friend Luigi Solari. It consisted of a glass tube with a plug of iron at one end and carbon at the other with a globule of mercury between. (The device is particularly interesting historically; it appears to be one of the forerunners of the semiconductor rectifiers used nearly a half century later.)

Marconi listened intently, growing more discouraged by the minute. Then, at 12:30 P.M., he heard the signals! Uncertain at first, he continued to listen. Soon there was no doubt. The faint but unmistakable signals were there, and Kemp was shortly to confirm their presence. The series of dots could be coming only from Poldhu, some 2,000 miles away. Marconi's second assistant, Paget, was ill on December 12, so was not present; he was to regret that illness all the rest of his life.

Faint signals were heard for a brief period on the next day, in spite of a howling storm. By December 14, it was apparent that obtaining evidence on the inker was not feasible with the equipment at hand, and that a better antenna could not be erected because of the terrible weather conditions. Marconi then had to decide whether to announce the results to the world. After all, there were no impartial witnesses to the accomplishment. Marconi cabled the results to London, then, on December 16, advised the press.

Immediately several stormy controversies arose. On December 16, the Anglo-American Telegraph Company which had a monopoly on message-carrying activities in Newfoundland, threatened legal action if the experiments were not terminated at once. Marconi decided not to contest the action. There were no such monopolies in the United States or Canada, and since he had laid out no significant amount for equipment in Newfoundland, it was easier to move than to fight.

The second and more far-reaching controversy involved the accuracy of Marconi's report. Many prominent scientists expressed doubt about what he had actually heard, believing that what he thought were signals could have been heavy static.

The *Western Electrician* of December 21, 1901, reported (in part):

> It is reported from Orange, N.J., that Thomas A. Edison, without casting any aspersion on Mr. Marconi, is doubtful of the reliability of the published statements. The newspapers thus report Mr. Edison: "I do not believe that Marconi has succeeded as yet. If it were true that he had accomplished his object I believe he would announce it himself over his own signature." When informed that Mr. Edison discredited the announcement of signals

having been received at St. John's last week from Cornwall, Mr. Marconi replied that the signals were received by himself, and that they were absolutely genuine. Further, he said, Governor Boyle, at Marconi's request, had cabled the fact to King Edward.

On the other hand, Professor M.I. Pupin of Columbia University is said to give full credence to the report. He is thus reported: "The signals were very faint, as I read in the report, but that has little to do with it. The distance, which is about 1,800 miles between these two points, was overcome, and further development of the sending instruments is all that is required."

A number of Chicago electrical men, such as B.J. Arnold, B.E. Sunny, E.J. Nally and E.B. Ellicott, were interviewed: they all united in substantially the same opinion—that the news was wonderful if true, but that further details must be awaited before final judgment could be rendered.

To this day, a number of responsible and respected scientists believe that no signals were actually heard on December 12, 1901, and that the story is a myth. Several reasons have been given to support this contention, the most important being the primitive nature of the receiving equipment used and the wavelength of the signal. Although the exact wavelength was not measured, Marconi has stated that it was about 366 meters (820 kHz).

In light of our present knowledge of propagation, the tests took place at the worst possible time of day for that wavelength, with both transmission and receiving site in daylight. There is little possibility that the frequency of 820 kHz could propagate over 2,000 miles in daylight. Absorption by the ionosphere is then at a maximum, and even powers of thousands of kilowatts would not deliver a significant signal over the Atlantic in that frequency range.

It has been theorized that the signals were not received at the frequency for which the equipment was designed, but at much higher ones. The Poldhu transmitter probably radiated many higher-order harmonics (frequencies twice, three, or even many times that of the fundamental frequency). Some of these could propagate over a daylight path, since daylight absorption decreases as the frequency increases. Radio amateurs of today frequently signal across the Atlantic with powers of only a few watts. Marconi may therefore actually have heard that historic series of dots on that bleak and dreary day, but on a frequency in the 10,000- to 20,000-kHz range, in what is now called the shortwave portion of the radio frequency spectrum.

In any case, doubts were dispelled less than three months later, when Marconi sailed across the Atlantic from Southampton to New York aboard the liner *Philadelphia*. The ship was equipped with the latest syntonic receiving equipment and an antenna on a specially constructed 150-foot mast. A Morse inker recorded the signals as they were received, and the captain of the ship verified all observations.

As the ship sailed farther from England, signals continued loud and clear. Up to a distance of 700 miles, they were recorded in broad daylight. Beyond that, the Poldhu transmitter could be heard only at night. This was the first observation of the nighttime effect that radio amateurs were to observe later—that signals (on certain frequencies, including the 200-meter band first allotted to amateurs) traveled much greater distances by night than by day.

During most of Marconi's historic voyage, signals were received before witnesses. Poldhu's S's were recorded to a distance of 2,099 miles, almost precisely the distance between Poldhu and St. John's. There could no longer be any doubt. The miracle of long-distance communication without wires had come to pass. Much of the miracle lay in the fact that this was only the beginning!

2

In the Beginning

Wireless did not spring full grown from the brain of Marconi, as many people suppose. It was instead the logical culmination of a long series of achievements by many scientists and inventors. Developments in electricity and magnetism date back to Thales of Miletus, who more than twenty-five hundred years ago observed that a piece of amber, rubbed with cloth, would attract certain light objects. (The word *electricity* itself is from the Greek word for amber, *electron.*)

Although the bridge across the ages from Thales to Marconi is marked by many notable achievements, actual radio communication is the work of a few brilliant men, whose labors can be traced from the first half of the nineteenth century.

Michael Faraday, one of the great original thinkers of all time, was one of these men. He was born on September 22, 1791, in what is now part of South London, the third son of a blacksmith. Michael's education was rudimentary, and at the age of thirteen he was apprenticed to a bookseller and bookbinder. Faraday was an avid reader; in particular, books on science fascinated him. He attended Sir Humphry Davy's scientific lectures at the Royal Institute in London, illustrating and binding the notes he made. In 1812 he sent the bound notes to Sir Humphry, with a request for a position in his laboratory. He was accepted and one year later was working at the Royal Institute as Davy's experimental assistant.

Faraday was soon doing original work in chemistry. In 1825, one year after he had been elected to the Royal Society, he isolated benzene, an achievement considered of great importance. Shortly thereafter he became director of the laboratory at the Royal Institute.

Faraday's electrical experiments had begun in 1820. By 1831 he was able to demonstrate that electricity could be produced by magnetism, a discovery that opened the way for large-scale production of electricity with alternating and direct current generators. Faraday's discovery of electromagnetic induction—he called it magnetic electricity—was a scientific milestone, for it led not only to a revolution in the production of electricity, but to one of the great theoretical achievements in the history of mathematics.

To explain how the magnet could act through space on the wire, most of the theoreticians of the time thought that "action at a distance" was taking place. Faraday thought otherwise. He postulated the existence of "lines of force" which spread out in all directions to fill space, and affected matter within that space.

Faraday's conception of these lines of force that permeated space was fundamental to the efforts of James Clerk Maxwell, who formulated a set of equations based on Faraday's "lines." Maxwell was a brilliant scientist. During his lifetime he made major contributions in astronomy, thermodynamics, and optics. Intrigued by Faraday's work and the concept of lines of force, he set for himself the task of expressing Faraday's ideas mathematically. Like Faraday, he disliked the concept of action at a distance through some undiscovered medium. Instead he set up a mathematical model to illustrate electromagnetic induction, in which changes in a magnetic field produced an electric force. Studying the equations, Maxwell realized that the process was reversible—that changes in the electric force would produce a magnetic field.

The laws of his system were expressed in a series of equations. The two primary Maxwell equations, from which the others can be derived, show that a changing electric field produces a magnetic field (as does a current) and that a changing magnetic field produces an electric field. Maxwell concluded that changing electric and magnetic fields in space would produce electromagnetic waves. He further asserted that these waves traveled through space with the velocity of light—that light itself was similarly produced by changing electromagnetic fields, and that light differed from other forms of electromagnetic radiation only in its number of vibrations per second (frequency).

Many scientists of Maxwell's time were reluctant to accept these theories, which so challenged the accepted ideas of the day. Maxwell's theory was completed in 1873, in a work entitled "A Treatise on Electricity and Magnetism." It would be fifteen years before a brilliant young German scientist, Heinrich Hertz, would prove the existence of Maxwell's electromagnetic waves.

Born in Hamburg in February 1857, Hertz chose a career in engineering at an early age, but changed his mind while in the army, pursuing instead a career in natural science. In 1878 he went to Berlin to study, and while there worked under two of the great scientists of the day, Hermann von Helmholtz and Gustav Robert Kirchhoff. Helmholtz quickly saw that the young Hertz had extraordinary scientific talents, and called his attention to a prize for research in electromagnetics offered by the Berlin Academy of Science. Von Helmholtz was one of the first scientists in Europe to give serious consideration to Maxwell's theory, and he encouraged Hertz to pursue the investigation of physical proof of that theory.

The apparatus with which Hertz accomplished that objective is shown in Figure 7. Not only was he able to show with that equipment that Maxwell's waves actually existed, he also showed that they could be refracted, reflected, and diffracted like light. Furthermore, Hertz showed that these waves traveled with the speed of light, just as Maxwell had predicted they would.

Hertz's apparatus consisted of an induction coil connected to a pair of metal rods. On one end of each rod was a flat metal plate, on the other a metal ball. The two balls were separated by a small adjustable gap. When the switch was closed, current from the battery flowed through the induction coil *primary*, a coil of a moderate number of turns of wire wound around a cylindrical iron (magnetic) core. The current was interrupted by a device like that on an old-fashioned doorbell. The changing magnetic fields set up by this varying current produced a current in the *secondary*, wound with many more turns. Because the voltage in an induction coil or transformer is proportional to the ratio of the number of turns in the primary and secondary, this current was produced at a very high voltage.

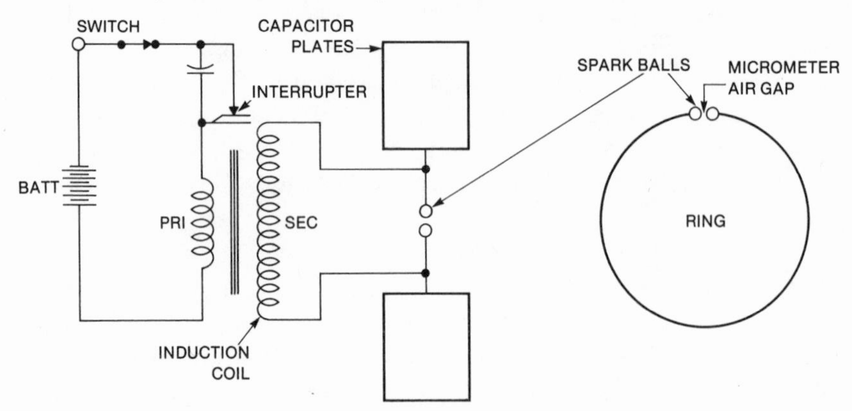

Fig. 7. The first radio transmitting apparatus—designed and used by Heinrich Hertz. *From* Encyclopaedia Britannica.

Fig. 8. Hertz's transmitter, left, and receiver, right, 1886–88. (Frequency was ultrahigh—about 250 MHz.) *Courtesy of the Deutsches Museum, Munich.*

The metal plates at the ends of the rods served as a capacitor (a storage device for electricity) which, when the stored energy was great enough, discharged across the gap between the metal balls. This discharge was *oscillatory* (what appeared to the eye as one spark was actually a number of sparks leaping back and forth with decreasing strength until they were too weak to jump the gap), and produced electromagnetic waves. Hertz detected these waves with a circular loop of wire, terminated by two metal balls with a minute gap between them. When the transmitter discharged through the spark gap, small sparks could be seen between the two balls in the receiving loop.

Hertz also determined—further confirming Maxwell's theory—that the waves were transverse (that is, that the electric and magnetic components of the wave were perpendicular to the direction of motion of the wave) by positioning his receiving loop at different angles and in different locations. It would spark when both transmitter gap and receiving loop were vertical, or both horizontal. When one was vertical and the other horizontal, no receiver spark appeared.

Hertz also proved—by hanging a large sheet of metal on the wall op-

Fig. 9. Some of Hertz's original equipment. At left, the polarized grid used to show the transverse nature of the radiation. At right is the 30-degree prism of pitch used to demonstrate their refraction. *Courtesy of the Deutsches Museum, Munich.*

posite his transmitter—that the radiation propagated as waves. He reasoned that metal would reflect waves. If the radiation did come in waves, the reflected and direct waves would cancel each other in certain locations and reinforce each other in other locations. And he found that at certain points between the transmitter and reflector no sparks appeared in the receiver.

By measuring the distances between the locations where the loop sparked and where it did not, and by calculating the frequency of the waves, Hertz was able to show that the radiation traveled with the speed of light. Both the existence of interference areas and the velocity of the radiation further confirmed Maxwell's theory.

Hertz published his findings in 1888, and Hertzian wave experiments were repeated in many laboratories. Four years elapsed, however, before anyone suggested that they might be used to communicate through space. Writing in the London *Fortnightly Review*, Sir William Crookes suggested that once a simpler and more certain method of generating waves of any desired wavelength could be found, together with more sensitive receivers that could respond selectively to desired wavelengths, it might be possible to communicate with them. (He also suggested directional antennas, or "means of darting the sheaf of rays in any desired direction, whether by lenses or reflectors.") Two friends "could thus communicate . . . by timing the impulses to produce long and short intervals in the ordinary Morse Code."

In spite of Crookes's amazing forecast, wireless communication "in the ordinary Morse code" was still far in the future. The means of using Hertzian waves for communication were not yet at hand. Hertz's oscillator was weak; it would send his waves only a few yards, hardly enough for a communications system. And the receiver was insensitive. Something better was needed, and would appear shortly.

Edouard Branly, a French physicist born in 1844, had become interested in physiology, particularly the study of nerves, which he observed to consist of large masses of closely packed fibers. Conducting experiments in which he attempted to duplicate the functions of these fibers, he observed that the resistance of a tube filled with fine particles of metal changed when subjected to either alternating or direct currents.

Branly next found that the tube's resistance also dropped in the vicinity of a lightning discharge. He then placed his tube in a circuit containing a battery and a galvanometer, and found that no current flowed in it. But when he placed an induction coil spark generator at distances up to 35 feet from the circuit, and produced sparks with his generator, he observed that current flowed in the circuit, as indicated by the galvanometer. He also observed that when he tapped the tube, the particles were somehow disrupted, and the current stopped.

Branly reported his findings to the French Academy in 1891, but attached no particular significance to the phenomenon. It was not until 1894 that Sir Oliver Lodge demonstrated it as a detector of Hertzian waves. He named Branly's device a *coherer* because the metal particles cohered under the influence of the waves, and substituted it for the receiving loop of Hertz's experiments.

Born in 1851, Lodge was appointed professor of physics at the University College, London, in 1881. He remained there during the years of his most significant contributions to the study of Hertzian waves, including tuning, the patents for which Marconi bought in 1911.

Alexander Stepanovitch Popov, Russia's wireless pioneer, was a lecturer in physics at the Russian Imperial Navy Torpedo School near St. Petersburg. Popov conducted experiments in which he attempted to develop a device that would detect thunderstorms in advance. In these experiments he used an elevated aerial—or antenna—to collect his thunderstorm signals, and a Branly coherer to detect them. He used a tapper, or decoherer, to restore his coherer after each discharge, and may have been the first to do so, though Lodge and Branly may have made the same discovery independently. Popov is believed by many to have made the first experiments in wireless signal transmission, using a Hertzian oscillator and a Branly coherer with his own decoherer. In March 1896 he transmitted the words "Heinrich Hertz" 250 meters "beyond the [St. Petersburg] University Botanic Garden." The coherer operated a Morse inker and recorded the dots and dashes on tape. The Russians consider him the inventor of radio communication, and May 7 (the date of an earlier and less well-authenticated transmission) is observed as Radio Day in the Soviet Union. However, since Popov did not take out patents on his invention, historians of science tend to credit the Italian Marconi as the man who not only assembled the components but brought the device to the attention of the world and established it as a practical means of communication.

In wireless, as in other fields of technological development, theory and practice have often marched together, sometimes with one ahead and then the other. Long before Crookes made his prediction, an American dentist, Mahlon Loomis, had seen the possibilities of wireless electric communication and in 1873 had formed the Loomis Aerial Telegraph Company. Loomis simply erected two "aerials" of about the same length (Figure 10), and made or broke a connection to earth in one of them, noting a deflection of a galvanometer in the other. He believed he was tapping a great electrical sea in the upper atmosphere, and by making and breaking the connection to earth was setting up waves in it that carried to his second aerial. He did succeed in detecting signals across thirteen miles of space, and experimented enough to learn that the "electrical sea" was variable and not always dependable. Financial problems cut his experiments short before the new company could find out just how practical the idea might be.

In 1875 Thomas Edison announced the discovery of a new "etheric force," which could not be controlled by insulation. This force was due to

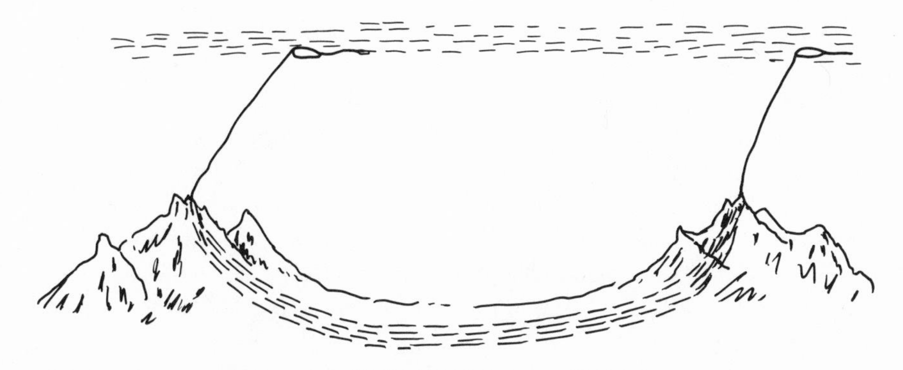

Fig. 10. How Mahlon Loomis believed his apparatus operated. He was the first person to propose electrical communication through space; first to conceive of doing so with the aid of an electric field (he called it a "sea"); and first to use an elevated conductor, or "aerial." *Reproduced by courtesy of the Library of Congress.*

the sparks from a magnetic vibrator, similar to the interrupter of an induction coil, and a wire attached to it would draw sparks from metallic objects. He turned the wire back on itself in a ring (something that Hertz was to do years later), and noted a spark as it touched any part of itself. This convinced him that he had discovered a previously unknown force. Applying several tests for (direct current) electricity, he decided that the force was not electric.

Elihu Thomson and David Houston, teachers in Philadelphia Central High School, had noted some interesting effects with a spark coil a few years before and, when Edison's report was published, repeated the experiments to test Edison's theory. With one end of the spark coil secondary attached to a large metal vessel and the other grounded, they were able to draw sparks from metal objects in the room and even to light a gas jet with a pointed finger. Thomson found he could draw sparks from the doorknob on each succeeding floor as he made his way upstairs to the library, six floors above. This—eleven years before Popov—was probably the first conscious transmission and reception of radio.

For Thomson and Houston believed the effect was electrical, and explained why Edison's tests were negative (why the effect would not register on a meter, for example). Each electrical impulse that produced a spark, they explained, was succeeded by a reverse impulse, an "inverse current," as they called it, which neutralized the effect of the first. (They did not know that there was a series of gradually weakening forward and "inverse" currents, or they would have had a complete and correct explanation of the oscillatory nature of the signal produced by a spark transmitter.)

Edison apparently was convinced, and in 1885 developed a wireless telegraphy system (afterward using it in his "grasshopper telegraph" from moving trains to telegraph wires at trackside. He spoke of his devices as working by "electrostatic induction." At that time, three years before Hertz announced the results of his experiments, no one knew of the possibilities of electromagnetic radiation; *induction* meant *electrostatic* induction unless otherwise stated. Edison's use of the word in that sense has led many modern readers, to whom *induction* and *magnetic induction* are synonymous, to misunderstand his statement, and dismiss his system as an "induction telegraph."

The confusion between induction—electric or magnetic—and true radiation is further illustrated by the statements of Professor Amos Dolbear, of Tufts College, Boston, Massachusetts. He gave what was probably the first demonstration of wireless telephony before the Society of Telegraph Engineers and Electricians, in London, England, March 23, 1882. His telephony range was limited to a few feet—from one room to another. He noted that "louder and better" signals were received when telegraphy, with a spark coil and an automatic interrupter, was used. With his spark equipment, he was able to transmit to any part of the building housing the Electrical Exhibition in Philadelphia in 1884, and even claimed a transmission range of 13 miles. He patented his circuit, but went no further with it, possibly in part because of crude receiving equipment, possibly because he, like many scholars, was more interested in research than communication. Like earlier experimenters, he did not at the time fully understand the phenomena he was producing. After the work of Maxwell and Hertz, most experimenters in the field were able to proceed more intelligently.

David Edward Hughes (1831–1900) discovered in 1879 that "an interrupted current in any coil gave out such intense extra currents that the whole atmosphere, even several rooms away, would have a momentary extra charge which was received by my telephones, even through such obstacles as walls." He used a loose contact between two pieces of carbon (the Hughes microphone) as a detector. He discovered that a loose contact between two pieces of metal would have the same effect, but that the pieces cohered after receiving a signal. He therefore abandoned the coherer fifteen years before Branly invented it.

Staging a demonstration before the Royal Society, in 1880 or 1881, he

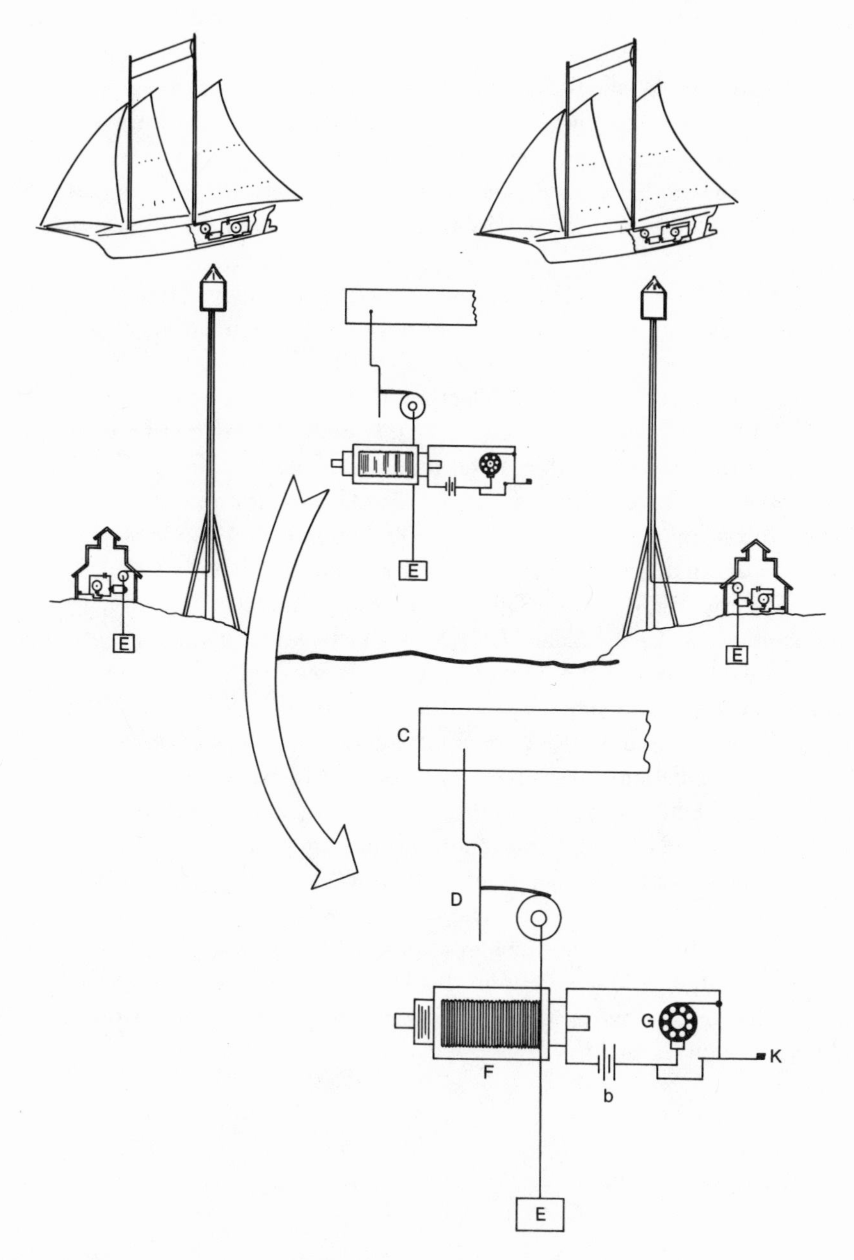

Fig. 11. An illustration from Edison's Patent 465,971, applied for in 1885, issued December 29, 1891. *Reproduced from Edison's Patent No. 465,971.*

transmitted and received signals over a distance of 125 feet. The members of the society, led by the secretary, Professor Stokes, insisted that the effects were due to simple (magnetic) induction. Hughes disagreed, and withdrew into a silence he was not to break for many years. However, persons who were later active in wireless saw and remembered the event. Hughes seems to have the distinction of having discovered Hertzian waves before Hertz, the coherer before Branly, and wireless telegraphy before Marconi.

In the next few years, several investigators demonstrated electric waves. Strangely enough, there was little interest in using them as a means of communication; even remote control received more attention. Jagadis Chundar Bose, of Bengal, in a lecture in Calcutta in 1895, transmitted waves 75 feet to another room of the hall, tripping a relay that fired a pistol and detonated a small mortar. Even the great Tesla, who lectured on radio transmission with tuned circuits at the Franklin Institute in 1893, was also interested in remote control, and in 1899 demonstrated wireless control of model boats in a large tank in New York's Madison Square Garden.

Augusto Righi (born 1850) was an early student of electromagnetic phenomena. A quiet researcher with no taste for publicity, he was not widely known even in the Italy of his time. It was probably his lectures on electromagnetic phenomena at the University that first interested the young Marconi in radio. Righi devised ingenious devices to help his study of electric oscillations. His spark gap was used by Marconi in his first experiments. Righi also devised a detector that was more sensitive than the earlier ones used by Hertz and others, by cutting thin lines in the silver on the back of a mirror. The gap thus attained was narrower than could be obtained with Hertz's micrometer gap.

One of Righi's greater achievements was to inspire the young Marconi. Not only did Marconi attend his lectures, but it is believed that the article that made him decide to make wireless a career was an obituary of Hertz, written by Righi.

3

In Search of a Detector

In spite of Marconi's remarkable achievements, his infant company was beset by difficulties almost from its inception. It was engaged, in 1902, in a series of bitter struggles to survive. Competition came not only from the established cable companies, but from rival wireless telegraph operations. In particular, the Germans, under the guiding genius of such men as Ferdinand Braun, Adolph Slaby, and Count von Arco, and the Americans, notably Reginald Fessenden and Lee de Forest, quickly set up competing services. Both the Germans and the earlier American systems were based on spark-gap transmitters.

To add to Marconi's difficulties, he had to spend much of his time and energy on costly, involved, and often bitter patent infringement suits, some of which continued for decades before they were finally resolved. Another factor that may ultimately have had an adverse effect on the interests of the company was the fact that Marconi was not convinced that wireless telephony had a future. He correctly foresaw the great importance of wireless telegraphy in ship-to-ship and ship-to-shore communications. When one of his assistants, Henry J. Round, developed a wireless telephone set during the 1906–1908 period, Marconi encouraged the project, but continued his primary emphasis on telegraphy.

Spark transmitters were not suited to the transmission of speech; spark transmission was not continuous. Each spark discharge resulted in a series of low-frequency oscillations, the first one of which was the strongest, followed by others increasingly weaker. The transmitted signal was thus continuously interrupted, resulting in poor articulation and intelligibility. Neither was the coherer—until about 1902 almost the only detector of wireless waves—usable for telephone reception. It was noncontinuous, requiring an automatic tapper to decohere the metal particles after each wave train had passed.

The coherer had other shortcomings, not the least of which was its slowness. With this and other factors on his mind, Marconi decided, on his return from the United States in 1902, to make a major effort to develop an improved way of detecting wireless waves. A better detector, he believed, would enable the Marconi Company to maintain its narrow lead over competing wireless telegraph companies. The result was the magnetic detector, or "Maggie," which Marconi himself developed. Joseph Henry had observed some sixty years before that rapidly alternating currents affect the magnetism in a steel or iron bar. Ernest Rutherford used a magnetic device experimentally in 1895 to show the presence of radio-frequency waves, and in 1897 Professor E. Wilson had constructed a wireless detector based on the principle.

As shown in Figure 13, the magnetic detector consists of a band of fine insulated iron wire that moves through a coil wound around a glass tube. One end of the coil is connected to the antenna, the other to the ground. Another coil of wire is wound around the center of the first one, and its ends are connected to the telephone receiver. Two horseshoe magnets,

Fig. 12. Early experimental apparatus, showing cohererlike and other devices. About 1899. *Courtesy of the Science Museum, London.*

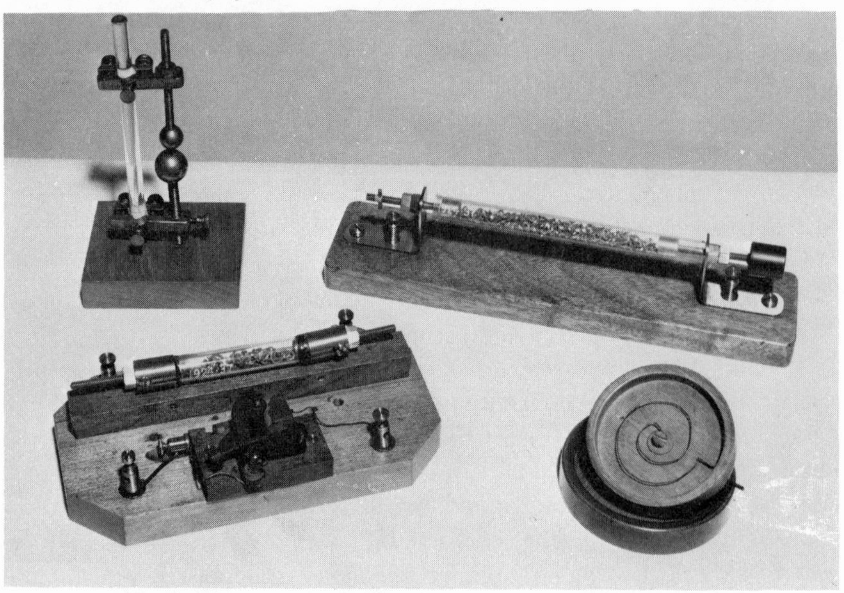

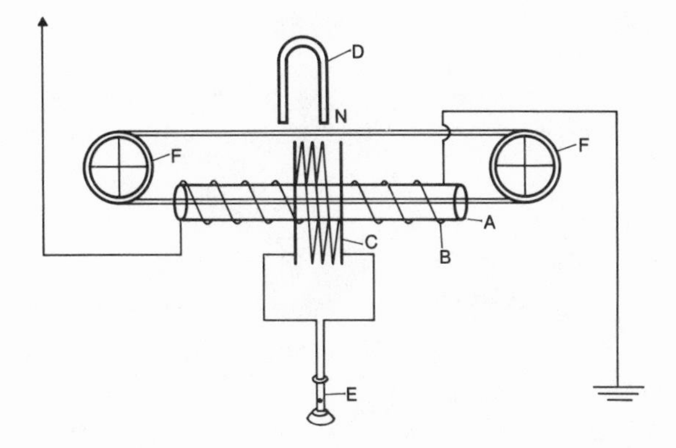

Fig. 13. Marconi's magnetic detector. Usually two magnets were used with their N poles together at the center of the coil. *From* Operator's Wireless Telegraph and Telephone Handbook, *by V. H. Laughter. Frederick J. Drake and Co., 1909.*

with their like poles at the center of the coil and their other ends beyond the coil, keep the wire magnetized. Because iron or steel wire does not lose its magnetization immediately after leaving a magnetic field, the magnetic effect is "dragged" along by the moving wire. A radio-frequency signal neutralizes this lagging or "hysteresis" effect, and the point of strongest magnetization slips back to a point directly under the pole-pieces of the magnet. This movement of the magnetism in the wires induces a current in the coil to which the telephone is connected, and sound is induced in the phones.

The "Maggie," as the new detector was called, was a decided improvement over the coherer, and was standard Marconi equipment for some years. Though not more sensitive than the coherer, it was rugged and reliable, and was much faster.

Meanwhile a similar search for a better detector was going on in the United States. Reginald Aubrey Fessenden, a Canadian who came to the United States in 1886 to work for Thomas Edison, was intrigued with the possibilities of wireless telephone communication. He taught electrical engineering at Purdue and the University of Pittsburgh from 1893 to 1900, then worked with the U.S. Weather Bureau until 1902.

In 1901, in connection with a not particularly successful demonstration of the transmission of voice via spark, Fessenden introduced a new type of detector. The device, which he called a *liquid barreter*, consisted of a silver-coated platinum wire immersed in a solution of nitric or sulphuric acid. The acid dissolved the silver, leaving a very fine contact between the

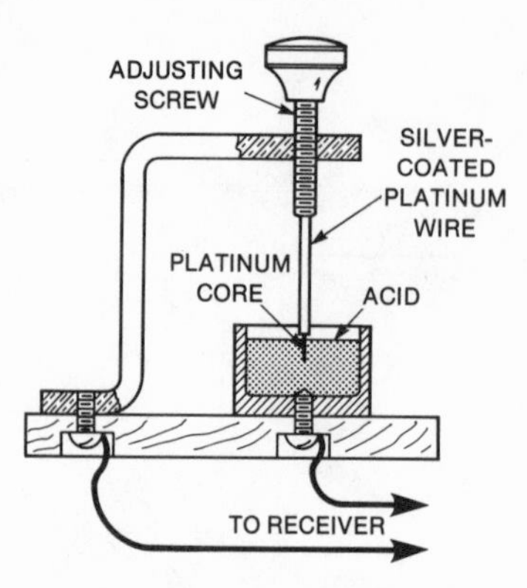

Fig. 14. Fessenden's electrolytic detector. The receiver mentioned consisted simply of a battery and a telephone receiver in series. The aerial was connected to the wire, the ground to the cup holding the acid. *Courtesy* CQ *Magazine.*

acid and the platinum. A battery connected between the wire and the acid caused a current to flow in the detector. This current was soon interrupted by the formation of tiny gas bubbles over the platinum. Wireless waves striking the detector ruptured the bubbles, restoring current flow. This flow was proportional to the strength of the incoming waves. The device, therefore, besides being more sensitive than the coherer, could receive continually changing wireless waves, and could be useful in telephony as well as in telegraphy.

Shortly after developing this *electrolytic detector,* Fessenden left the Weather Bureau to form his own organization, the National Electric Signalling Company, with the backing of two Pittsburgh financiers, Hay Walker, Jr., and Thomas H. Given.

In 1903, Greenleaf Whittier Pickard, grandnephew of the poet John Greenleaf Whittier, found that some minerals, when in contact with a metal, are relatively good conductors in one direction, and poor ones in the other. This made the mineral-to-crystal contact act like a mechanical valve, conducting current in one direction, but blocking it in the other. He immediately envisioned the use of these minerals in wireless receiving circuits, because he knew that the high-frequency wireless currents, when rectified, could be heard in telephones.

Pickard discovered that some crystalline forms of matter were especially useful as rectifiers. The crystals were held in a metal cup, usually embedded in a metal alloy (Wood's metal) having a low melting point and were contacted by a thin wire that barely touched the surface of the crystal. He found that certain points on the material were much more sensitive than others. The chief disadvantage of these delicate wire—or "cat's whisker"—contacts was that they were not rugged, and could be displaced by vibration of any kind. This made them unsuitable for use aboard ship, where heavy seas could disturb the delicate balance on the surface of the crystal.

Pickard experimented with hundreds of different crystals, including galena. His most effective detector was silicon, which he first used in 1906. Silicon crystals were stable and more uniformly sensitive. Possibly his most important contribution was the "Perikon" detector, two minerals—a copper ore called chalcopyrite, and zinc oxide—held firmly together. Pickard's work with mineral rectifiers foreshadowed the development of the transistor forty years later.

In 1906 General Henry Dunwoody of the U.S. Army developed the carborundum detector. Not quite as sensitive as galena or silicon, it abandoned the light contact of the cat's whisker for a firmly embedded steel point, making it useful aboard ships, where rolling seas jarred the cat's whiskers off the sensitive spot on the commonly used galena crystal.

The crystal detector was quickly and universally adopted. It was inexpensive, easy to use, and sensitive. For some ten years it was almost the only detector used by amateurs, and was a major factor in the growth of amateur radio during the 1900s.

Meanwhile, de Forest was experimenting with a long series of flame detectors, one of which, in highly modified form, was to be the solution to the problem of the better detector, and somewhat later was to solve the problem of a better transmitter as well.

Marconi's triumphs had made *wireless* a household word. The popular press was fascinated by the new medium. Magazines and newspapers carried articles and stories about wireless, and printed instructions for building wireless sets. They could be constructed at a reasonable cost from readily available materials, and the romantic and exciting concept of transmitting intelligence over great distances fired the imagination of many, especially the young. These amateurs, mostly self-taught, without formal technical training, were to make significant contributions to the development of wireless, blazing new trails into regions hitherto untraveled by humankind.

4

Continuous Waves

Fessenden was convinced that he could design wireless equipment that would transmit the human voice across the Atlantic, and shortly after forming the National Electric Signalling Company, he began to construct a large station at Brant Rock, Massachusetts. The equipment included an improved rotary spark gap, which Nikola Tesla, American physicist and inventor born in what is now Yugoslavia, had patented in 1896. It was a metal motor-driven wheel with studs around its periphery. The wheel rotated between the electrodes of a spark gap, and the spark, which would normally have jumped from one electrode to the other, now jumped from one electrode to the stud passing it at the moment, and from the stud on the opposite side of the wheel to the other electrode. The advantage of the rotary gap was that the number of sparks per second could be varied to produce an easier to copy note. The pitch of the signal depended, of course, on the number of sparks per second, which in turn depended on the number of studs on the wheel and the speed at which it rotated. Wireless operators soon discovered that they could identify a rotary-gap station by the pitch of the note it emitted.

Wireless telephone transmissions with the rotary gap were better than with the older fixed spark, but still not quite good enough, and Fessenden decided that the only way of transmitting good-quality voice signals was by *continuous waves*. Some years earlier Nikola Tesla had experimented with continuous-wave generators. Fessenden concluded that he could design a machine that would produce continuous waves suitable for wireless telephony.

The common alternator—the machine that generates our alternating electric current—operates at a frequency of 60 hertz (cycles per second).

The current changes direction 120 times a second (a *cycle* consists of a pulse of current in one direction, then one in the other, along the wire or conductor, after which the *cycle* repeats). Radio waves are produced at much higher frequencies, from 20,000 up to millions and billions of hertz (Hz). Formidable problems faced the would-be developer of any device that could produce such frequencies. The number of poles in the generator would have to be increased greatly, as would the speed of rotation, since the frequency is the product of the number of pole-pairs and the rate of rotation, in revolutions per second. Tesla's 1890 generator had 284 poles and produced a current at 10,000 Hz.

Undaunted, Fessenden submitted specifications for such an alternator to the General Electric Company, which assigned the project to Charles Steinmetz, one of the world's greatest electrical engineers. In 1903, an alternator operating at 10,000 Hz at a power of 1 kilowatt (KW) was delivered, but Fessenden was not satisfied. He then ordered an alternator that could deliver 100,000 Hz.

The task of building the 100,000-Hz alternator was assigned to Ernst F.W. Alexanderson, a young Swedish engineer who had begun working for General Electric several years before. Alexanderson suggested some changes in the plans for the generator. Fessenden had specified a wooden armature; Alexanderson felt an iron armature would be better. Fessenden insisted that the alternator be constructed in accordance with his specifications, and a 50-kHz, 1-kW alternator was delivered to him in September 1906. Although it operated at a lower frequency than specified, he accepted it. Initial tests were unsuccessful, and the machine had to be taken apart, only to find that the trouble was a simple broken wire.

With this alternator, Fessenden made the first radiotelephone broadcast in history, from Brant Rock on Christmas Eve 1906. The program started with the general call to all stations, CQ, CQ, in Morse code. Operators sprang to attention at wireless stations along the Atlantic coast and aboard ships within a radius of a few hundred miles. Then to their astonishment, they heard a recording of Handel's *Largo*. Fessenden, the only one at Brant Rock who could play a musical instrument, followed with a solo on the violin. This established him as history's first live radio performer.

Reception reports flowed in from as far south as Norfolk, Virginia. The experiment was repeated the following week, and this time reports were received from as far away as the West Indies.

Fessenden's genius extended beyond efforts to generate continuous waves (CW) that would carry voice. He was also occupied with improving reception. His electrolytic detector had been a great advance on the coherer, but his most important contribution to receiving techniques was the *heterodyne receiving system* (Greek *heteros,* "other," and *dynamis,* "power" invented in 1905). Since few adult humans can hear frequencies above about 20 kHz, his continuous-wave transmissions, if broken up into the dots and dashes of the Morse code, would be inaudible. A listener with an ordinary detector would hear only a click for each dot, and a click at the beginning and end of each dash. (The Christmas Eve broadcast was made by varying the strength, or *amplitude,* of the signal with a microphone in the antenna circuit.) After the 50-kHz waves were rectified by the detector, what the listener heard was a direct current that increased or decreased in strength in exact proportion to the increases or decreases in antenna current caused by the microphone.

Heterodyning means combining two frequencies to produce a third. Fessenden reasoned that if a frequency of 98 kHz, for example, were being received, and another frequency, say of 100 kHz, were injected into the receiving circuit, two new frequencies, the sum and difference of 100 and 98 kHz, would be produced. The 198-kHz sum signal could be filtered out or ignored, and the 2-kHz difference signal would be heard in the phones. So visionary was the conception that no practical way of utilizing it had yet been developed, and it would be many years before it could be exploited practically and economically.

While Fessenden toiled to make his company the first transatlantic wireless telephone organization, other significant developments were taking place. Alexanderson, for example, continued his work on the alternator. The General Electric Company had taken out patents on the young Swede's ideas and directed him to continue work after the Fessenden alternator had been delivered in 1906. Working with the concept of an iron armature, Alexanderson patented a 100-kHz alternator in 1909. It had 200 pole-pairs on the stator, and the rotor speed was 30,000 revolutions per minute (rpm). Fessenden was present when it was first tested, and was so impressed that he ordered that an alternator he was having built by General Electric be modified to use an iron armature.

Meanwhile, other scientists were investigating methods of improving spark transmission, and to develop other ways of generating continuous waves. Still others were trying to find better detectors.

Around 1906, the Telefunken Company introduced a *quenched spark-gap* transmitter based on a discovery by German physicist Max Wien. The quenched gap was made up of a number of metal discs spaced close together, separated only by thin mica rings. Instead of jumping across a wide gap between balls or rounded-end electrodes, the spark made a number of jumps across the much shorter space between the discs. The lower voltage sparks died away more quickly and the gap became an open circuit almost immediately.

Why was this an advantage? It has been pointed out earlier that what appears to be a single spark in a Hertzian oscillator is in fact a series of sparks. When the current in the primary winding of the induction coil is interrupted, the magnetic field around the coil collapses, causing a surge of current to flow in the high-voltage secondary. This charges the capacitor (condenser in the days of spark) until it reaches a voltage that breaks down the resistance of the air in the spark gap, and a spark jumps. Immediately current rushes into the coil and builds up a magnetic field around it. As soon as the capacitor has discharged to where the spark is extinguished, and no more current is flowing into the coil, this magnetic field collapses, producing a current that charges the capacitor again, this time in the opposite direction though not to the same extent, because some of the electric energy has been turned into heat in the conductors and across the gap. The circuit is said to have *losses*.

If some way could be devised to take the spark gap out of the circuit immediately after the first discharge, the coil-capacitor in the antenna circuit could go on swinging back and forth, oscillating as just described, without being loaded down by the spark. Oscillation would be freer, the train of waves would be stronger and take longer to die away and would be less *damped*, and more of the transmitter's energy would be turned into radio waves instead of heat.

The quenched gap did just that. By quenching very quickly, it removed itself from the circuit and ceased to load it down. This resulted in a more continuous, more effective, and quieter operating transmitter.

But in spite of all the improvements, the days of spark were numbered. Another equally revolutionary method of generating continuous waves—the Poulsen arc—was to compete with the Alexanderson alternator. These two devices for producing continuous waves would control the field until the coming of that most momentous of all wireless inventions—the vacuum tube.

Valdemar Poulsen was born in Denmark in 1869. By the turn of the century he had distinguished himself by inventing the first magnetic recorder—the *Telegrafone*, which was the forerunner of all wire and tape recorders. Poulsen then devoted himself to methods of producing continuous waves. He turned his attention to the electric arc.

In 1900 the British scientist W. B. Duddell had produced a singing effect with an arc lamp. Even though the arc was energized by direct current, Duddell had been able to produce an alternating-current note at audio frequencies when he connected an inductance and a capacitance across the arc lamp. Poulsen was able to use the arc to generate radio waves at much higher frequencies.

The Poulsen arc consisted of a carbon and a copper rod. These *electrodes* could be brought together or separated to form a gap of any desired length. To start the arc, the electrodes were brought together and current turned on. The electrodes were then separated, "striking" the arc. The arc was enclosed in a chamber filled with hydrogen gas, which improved its operation by increasing the amount of ionization between the electrodes and thereby decreasing the arc resistance.

An arc, as well as a spark gap has *negative* resistance—if the voltage across it increases, the arc grows "fatter," more gas is ionized between the electrodes, and a lower-resistance path is provided for the electrons. If a coil and capacitor are placed across the arc, as in Figure 15, and the switch is closed while the arc is conducting, capacitor C charges. Since the arc and capacitor share the current, the current in the arc drops, and the voltage across it rises, causing more current to flow into the capacitor. Meanwhile a magnetic field is building up around the coil.

When the capacitor is fully charged, current ceases to flow through the coil, and the magnetic field around it collapses, pouring more current into the capacitor and raising its voltage above the supply voltage. It immediately starts to discharge through the arc, increasing its current and therefore lowering the voltage across it, thus increasing the capacitor discharge.

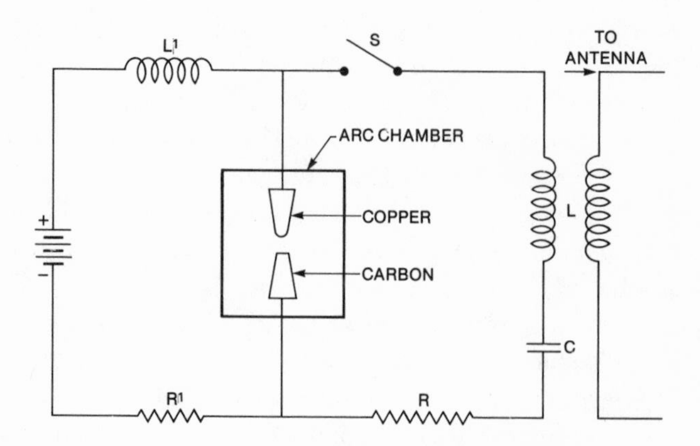

Fig. 15. Poulsen arc generator, schematic representation.

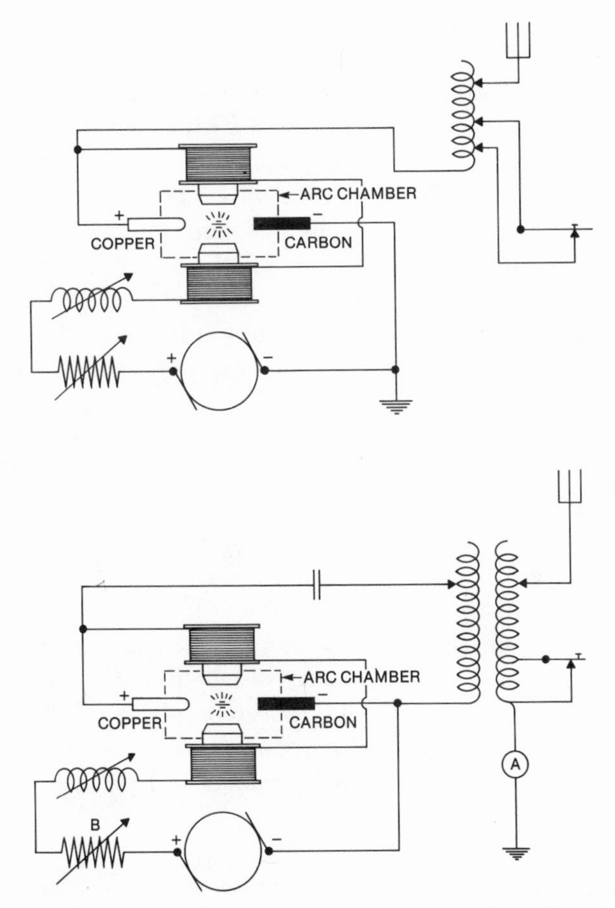

Fig. 16. Two diagrams of Poulsen-type arc transmitters. *From the 1921 U.S. Army Signal Corps manual.*

When the capacitor again charges, voltage across the arc starts to rise, and the cycle repeats. By adjusting the values of R, L, and C in the figure, continuous high-frequency radio waves were produced. The characteristics of the arc were such that frequencies above about 250 kHz were difficult to produce.

The arc was widely used in Europe, and in 1909 Poulsen was approached by a young electrical engineering graduate from Stanford University, Cyril F. Elwell, with a proposition to purchase the United States rights to the invention. Elwell, who had no experience in wireless and was working on electric furnaces, had been persuaded to test a spark radiophone. Like other spark telephony devices, it worked, but not well enough to produce intelligible speech. But Elwell noticed that it came nearest to being intelligible when the spark gap became so small as to be practically an arc. This interested him in Poulsen's device, and he went to Denmark to hear it. Convinced that it was excellent for telephone and much better for telegraphy than spark, he attempted to raise money to form a company to exploit the American rights. After a second trip to Denmark and the purchase of a small arc transmitter, he succeeded.

Elwell demonstrated the arc to Stanford University faculty members and others, and was able to form with their help the Poulsen Wireless Telephone and Telegraph Company (afterward Federal T&T). Stations were built in Stockton, Sacramento, and Palo Alto, California, and tests demonstrated not only that the new transmitters were faster in telegraph than spark, and transmitted voice as well as the line telephone, but that they were much more selective than spark, and that two transmitters on slightly different frequencies could transmit without interfering with each other. Stations were built in San Francisco and Los Angeles in 1911, and in 1912 the company had fourteen stations, ranging from Chicago to Honolulu, in operation. Honolulu was 2,400 miles from the San Francisco station, and two-way operation at night was reliable. Late in 1912 the Navy asked for tests, and a 30-kW arc was installed in the Navy's powerful Arlington, Virginia, station. The little arc exchanged messages with Honolulu, a distance of 4,500 miles and a record for the time. The Navy's 100-kW spark, at the same location, had never been able to reach its Mare Island station off California.

Properly impressed, the Navy ordered a 100-kW Federal T&T arc transmitter for Darien, Canal Zone, and followed it with orders for 350-kW transmitters at Pearl Harbor, Hawaii, and Cavite, in the Philippines. The United States navy thus became the first in the world to use continuous wave.

5

The Audion

Developments in wireless—as in other fields of technology—often overlapped each other, and sometimes occurred simultaneously. The invention of the vacuum tube not only overlapped the development of other types of detectors, but actually predated them.

In 1882, long before anyone thought of crystal, electrolytic, or magnetic detectors, Thomas Edison, working to perfect his incandescent lamp, was trying to discover the reason for the breakage of his lamp filaments and blackening of the bulbs. He covered the inside of one of the bulbs with tinfoil and connected the foil to the negative terminal of the lamp filament, inserting a galvanometer between the filament and the foil connection. Nothing happened. But when Edison connected the foil to the positive terminal of the filament, a small current was indicated by the galvanometer. (See Figure 17.) Edison recorded his observations of the phenomenon, now known as the "Edison effect," and in 1884 patented an electrical indicator based on the effect.

The Edison effect would prove to be one of the most important phenomena in the history of science, but because it appeared to be an abstract thing, with no practical application, little attention was paid to it. More than ten years would pass before John Ambrose Fleming would turn his attention to it, thereby foreshadowing for mankind a great leap forward in the development of communications technology.

Fleming, in 1904, was a technical consultant to the Marconi Company. He was searching for improved detection methods, and experimented with chemical rectifiers, without marked success. Then he recalled the Edison effect. While employed as an advisor to the Edison Electric Light Company of London in 1882, Fleming had worked on the black-

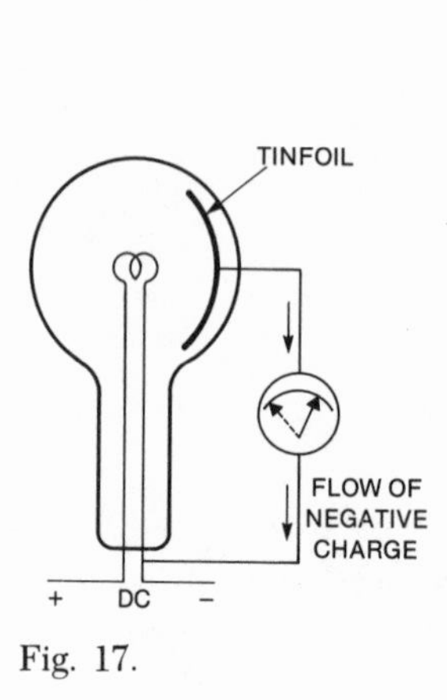

Fig. 17.

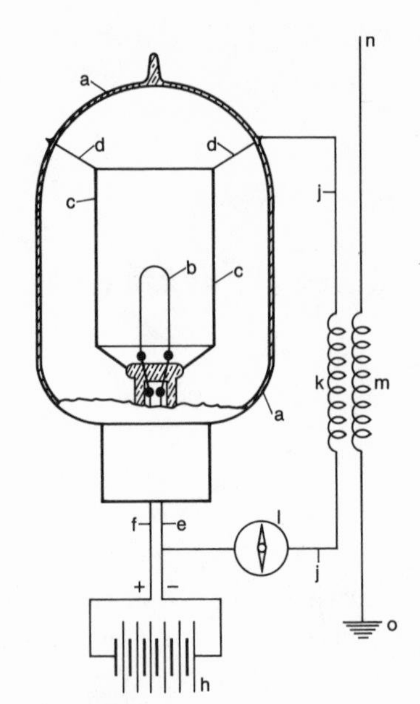

Fig. 18.

ening-bulb problem. Sidetracked at the time by other apparently more important work, he retained a number of the carbon-filament bulbs with which he had been working. Now, in 1904, he located one of the lamps and connected it across a direct current (DC) voltage. He then connected the enclosed tinfoil to a galvanometer and an inductance coil in series, and then to the negative terminal of the filament supply, as shown in Figure 18. Just as he had hoped, as the wireless waves induced a current in the coil, the galvanometer registered a flow of direct current. The current in the coil was alternating (AC), changing its direction at the frequency of the transmitting station. When the coil current made the tinfoil positive with respect to the filament, a current flowed. On the other half of the oscillation, when the current made the foil negative, nothing happened. Fleming had discovered—in his old light bulb—a method of rectifying high-frequency wireless waves. The diode detector was born!

Fleming immediately arranged for the construction of an improved "lamp," in which he replaced the tinfoil with a metal cylinder that surrounded the filament. It worked, as hoped, much better than the original bulb. Fleming called it an *oscillation valve,* because it acted as a one-way valve, permitting a flow in one direction, but none in the other. Replacing the galvanometer with a telephone made it the wireless wave detector he had sought. He applied for a patent in 1904. The Fleming valve, as it came to be known, gradually replaced the Maggie. Not quite as sensitive, in the minds of some old operators, it was much less affected by heavy static discharges, and was equally rugged and stable.

The three-element vacuum tube, or Audion, has been called one of the most important inventions in history. It developed into a practical amplifier, an extremely sensitive detector and an oscillator, and made possible all the giant strides made in the industry during the first half of the twentieth century. The inventor, Lee de Forest, was born in Iowa in 1873. His father was a Congregational minister. From boyhood the young de Forest wanted to be a scientist and inventor. In 1893 he entered the Sheffield Scientific School at Yale University—helped by a scholarship set up by a distant relative for young men named de Forest—and received his Ph.D. in 1899. His doctoral thesis was entitled "Reflection of Hertzian Waves from the Ends of Parallel Wires." Thus he became the first American to write a thesis on the infant wireless.

After graduation, de Forest went to work for the Western Electric Company in Chicago, earning $8 per week in the dynamo department. Wireless continued to fascinate him, and he was soon spending more and more of his time attempting to develop improved detection devices. His

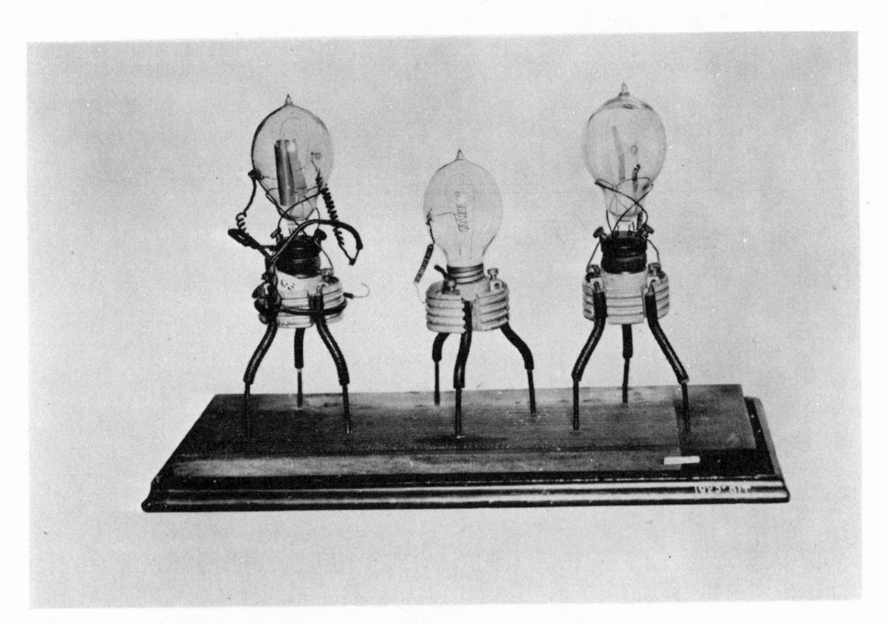

Fig. 19. *Courtesy of the Marconi Company.*

first attempt he called a *Responder*. Based on the experiments of the German Aschkinass, it was a liquid anticoherer. Tiny "bridges" in a drop of liquid between two metal electrodes carried current until a wireless wave was received. The waves broke down the "bridges" and raised the resistance of the device—an action exactly opposite to that of Branly's coherer. He and an associate applied for a patent on this detector in 1900.

In 1902 de Forest had formed the American de Forest Wireless Telegraph Company, and during the next four years took out some thirty patents, none of which was particularly important. The search for new and better means of transmitting and detecting wireless waves continued, however. One of his devices, an electrolytic detector, became the subject of complex, and sometimes bitter, patent infringement litigation.

In 1903, de Forest had visited Fessenden at his laboratory, and had seen the liquid barreter in action. The basic principle of Fessenden's patented detector was the use of a Wollaston wire, a thin platinum wire enclosed except for the very tip in a silver sheath. De Forest returned to his laboratory resolved to use a Wollaston-wire detector or its equivalent. Studying the history of the Wollaston wire, he found that Pupin had used a fine platinum wire, insulated except for its end, and immersed in an acid electrolyte, long before Fessenden. Following Pupin, he produced what he called a *spade rectifier*. A Vermont federal judge, however "could find no difference between the patented fine wire point dipping into an electrolyte and a fine end of one sealed into a glass insulator," found de Forest guilty of infringement, and fined him, in spite of the "previous disclosure" by Pupin. By this time—1906—crystal detectors were available, so the de Forest company's operations were in no way curtailed.

In 1899, it is said, de Forest had read an account by Tesla describing the possibility of using a flame between two electrodes as a detector of wireless waves. A detector using a flame had also been used by Rutherford. A platinum electrode is inserted into the base, or coolest part, of a Bunsen burner flame, and another into the tip, or hottest part. The tip electrode is connected to an antenna, the other to a ground.

How much de Forest may have known about these earlier experiments is not clear, but an accident impelled him to study flame detectors. He and an associate, Smyth, roomed together in Chicago. They noted that when they operated a spark coil in the room, the Welsbach (mantle-type) gas burner brightened. Further experiment showed that the effect was due to sound waves from the spark, but the idea of a gaseous detector remained in de Forest's head, and during the next few years he took out no less than five patents on different types of flame detectors, one of which had no less than four ring-shaped electrodes surrounding the central flame. All these detectors were hooked up as relays, in which the incoming wireless signal was to trigger or intensify the current supplied by a battery in the circuit. De Forest called this a "booster" or B-battery. In 1905, de Forest began to experiment with gas-filled tubes, and took out five more patents, pretty much repeating his earlier experiments with open flames.

Attempting to improve the sensitivity of his detector, de Forest wrapped a piece of tinfoil around the outside of the tube, connecting this electrode to the antenna. It then occurred to him, he said, "that the third, or control, electrode could be located more efficiently *between* the wing [plate] and filament. . . . I decided that the interposed third electrode would be better in the form of a grid, a simple piece of wire bent back and forth and located as close to the filament as possible.

"I now possessed the first three-electrode vacuum tube—the Audion, grandaddy of all the vast progeny of electronic tubes that have come into existence since."

There has been much controversy over the parts played by Fleming and de Forest in the development of the Audion. The simplest approach, used by many historians, is that "de Forest added a third electrode to the Fleming valve." De Forest himself stated that at the time he was quite unaware of the existence of Fleming's invention, and said: "During the time I was developing the two-electrode detector I had never heard of the Fleming valve, and was therefore surprised when I later learned that my invention was being confused with it."

One well-known writer states that "it would be naïve to suppose that Fleming's valve might not have given him a suggestion, . . . yet the distinguished British author and editor G.W.O. Howe says "De Forest claimed not to have known of the existence of the Fleming valve, and I see no reason not to believe him." A further interesting point is brought up by a book of the period, *Operators' Wireless Telegraph & Telephone Handbook*, by Victor H. Laughter, who at the time was a radio instructor (technical director, American Wireless Institute, Detroit. He later joined Hugo Gernsback as editor). Laughter described some ten "Detectors and Detecting Instruments" in Chapter VI of his book. These included the carborundum detector, which was invented in 1906, and de Forest's three-element Audion, but does not mention the Fleming valve. Apparently Laughter had not heard of it, yet the book must have been written no earlier than 1907. (Incidentally, Laughter says of the Audion: "It is doubtful if it will ever

Fig. 20. *Courtesy of Arthur Trauffer.*

come into wide use, owing to the difficulty in manufacture and short life.")

In view of the fundamental facts, the question of whether de Forest had ever seen or heard of a Fleming valve is irrelevant. For the two devices actually belong to two distinct *families* of detection devices, in spite of their superficial resemblance. The Audion, like de Forest's earliest Responder, was powered by a local battery, current from which flowed through the phones. The arrival of a wireless wave reduced or increased the current. Thus the de Forest devices, like Branly's coherer, were *relays*, in which small amounts of radio-frequency power controlled much larger amounts of power from a local source. The Fleming valve was a *rectifier*, like the Fessenden liquid barreter and the various crystals (silicon, germanium, and carborundum) and therefore worked on a different principle.

Further indications that de Forest was following his own course are given by the five patents he took out on flame-operated detectors—all of which used the B-battery—and the fact that his five patents on vacuum detectors appeared to be following the steps of the earlier patents on open-flame detectors.

The patent for the Audion was granted February 18, 1908 (Patent No. 879,532). An earlier patent on a three-element tube (No. 841,387), which embodied the principles of the flame detector with two plates on opposite sides of the flame, did not have a grid. The elements were a filament and two plates. It was, however, entitled: "Device for amplifying feeble electrical currents." These two patents have been said to be among the most valuable ever issued by the United States Patent Office.

The Audion was the "great leap forward" in the history of radio, for it offered a means of controlling and eventually amplifying weak incoming signals. The filament of the tube—also called the cathode—was heated by a battery. Another, booster battery kept the wing or plate positive with respect to the filament. The heating caused the filament to emit electrons. Without a grid, some electrons flowed toward the positive plate. With a grid near the cathode, the tube became not only an extremely sensitive detector but, when de Forest learned the value of a higher vacuum, an amplifier as well.

In practice the grid was connected to the antenna. As the voltage on the grid changed because of variations in the received signal, so too did its effect on the electrons in the space within the tube. When the voltage on the grid was negative, electrons were repelled back toward the filament; when the grid became positive, the grid attracted the electrons, and greatly increased the number flowing toward the plate. Thus the Audion acted as a rectifier. But it was more than a rectifier. Because a small change in the voltage on the grid resulted in a large change of plate current, the Audion not only detected weak signals, but amplified them as well. Early Audions amplified very little—it was not until 1912 that tubes pumped to a "harder" vacuum became practical amplifiers—but they were fantastically better detectors than anything that preceded them.

Although the American de Forest Wireless Telegraph Company had sold considerable equipment during the few years of its existence, it ran into financial difficulties in 1907. According to G.E.C. Wedlake in *SOS—The Story of Radio Communication,* the president of the company, Abraham White, "with the connivance of other directors, carried out a Stock Exchange transaction which converted the American de Forest Wireless Telegraph Co. into the United Wireless Co. of America." De Forest says, "[they] had transferred to it all the assets of the American de Forest Telegraph Co., leaving all its debts in the empty shell of the old company for the benefit of creditors!" Wedlake continues: "De Forest refused to have any part in the very dubious transaction and resigned from the company, being paid the ridiculous sum of $1,000 for his patents," though the directors graciously allowed him to keep his Audion and Aerophone (arc-transmitter) patents, considering that the patent applications in question had no value.

De Forest immediately formed a new company, the de Forest Radio Telephone Company, and continued to work on developing an efficient radio telephone. He used an arc transmitter of a type not covered by the

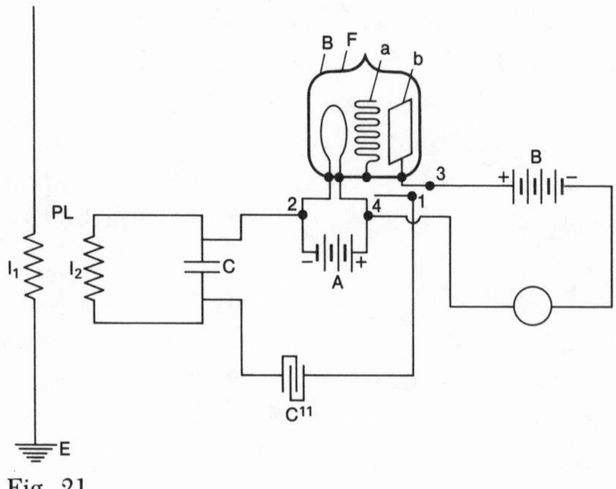

Fig. 21.

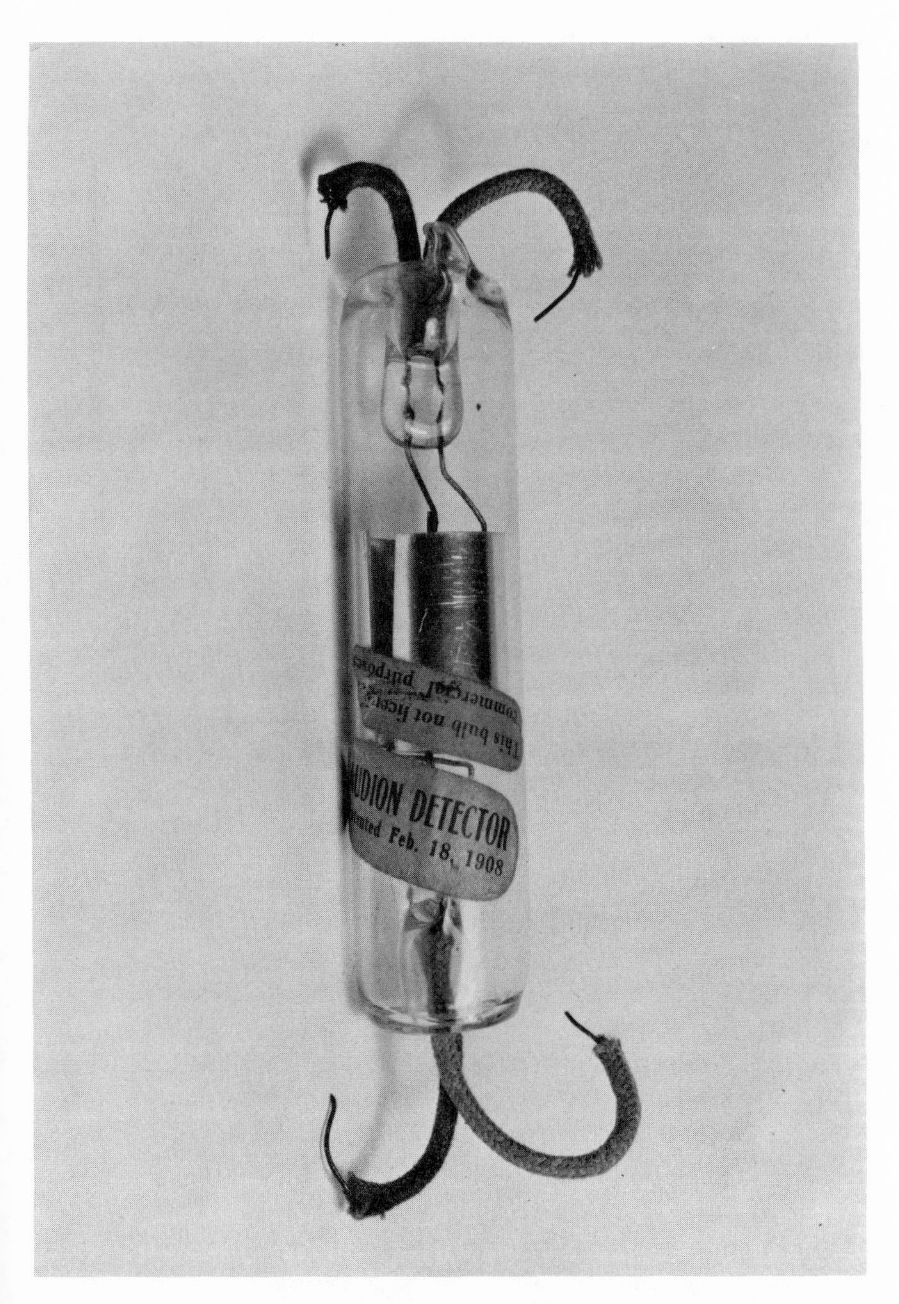

Fig. 22. An early Audion. *Courtesy of Arthur Trauffer.*

Poulsen patents, and employed both Audion and crystal detectors in his receiving equipment. In 1908 he journeyed to Europe with his wife to demonstrate and sell his equipment. While in Paris, he was given permission to conduct experimental tests from the top of the Eiffel Tower. De Forest and his wife played phonograph records all night, and the test transmissions were heard as far as Marseilles, 500 miles away.

One immediate result was an order from the Italian navy for four of his Aerophone arc installations. The United States navy had earlier bought a number of Aerophones, installing them on sixteen battleships, six destroyers, and two auxiliary vessels, for a round-the-world cruise. Hurriedly installed and manned largely by untrained personnel, de Forest felt that they gave "very poor results." The commander of the fleet, Admiral Evans, however, was enthusiastic and remained a convert to radio from that time on.

Meanwhile, in 1911, the de Forest Radio Telephone Company was again in serious financial difficulty. The president of the company had suddenly resigned not long before, after having, de Forest claimed, "gutted the treasury." The company fell on hard times, ameliorated from time to time by sales to the Army or Navy. Traveling to the West Coast to install equipment on two Army transports, and he met C. F. Elwell of Federal T & T. Since reports from the East were unfavorable, he suggested to Elwell that he might find it advisable to seek work on the West Coast. He immediately found himself in a position at Federal, soon to become its chief engineer. De Forest and his assistant, Charles Logwood, attempted to make an amplifier to strengthen the weaker signals received at the company's radio stations. At first they used mechanical relay-type amplifiers, then the Audion. Two Audions in series (cascade) did give some results, but as the voltage on the tubes was raised over about 20 volts, they "blue-hazed" and stopped working. A number were sent to an X-ray manufacturer in San Francisco, who pumped them to the same "hardness" as his X-ray tubes. To the delight of de Forest and Logwood, the "hard" tubes were able to work with plate voltages as high as 120, and the audio amplifier became an established fact.

While with Federal, de Forest was placed under arrest with some of the officers of his company, charged with using the mails to defraud in connection with attempts to sell stock in the de Forest Radio Telephone Company. He and his attorney Darby were acquitted, though three of the leading figures in the company received jail sentences.

6

Early Regulation

During the early years of the twentieth century, competition for the limited wireless business that was available was bitter. Telefunken in Germany, Ducretet in France, Fessenden's National Electric Signalling Company, and de Forest's American Wireless Telegraph Company in the United States were at a serious disadvantage because Marconi had been an astute businessman as well as an inventive genius. He took patents out quickly on all patentable inventions, and contested them in the courts where in his judgment they had been infringed upon. Marconi shore stations and Marconi-equipped ships were not allowed to handle traffic from foreign installations, with the important exception of distress calls. In addition, Marconi refused to sell equipment outright, but insisted on a leasing arrangement which allowed the user a certain number of words per month as part of the arrangement, with a surcharge based on the number of words above the minimum. At the earliest opportunity, Marconi organized an American Marconi Company, chartered in New Jersey in 1899, and quickly seized an opportunity that availed itself to sell equipment to the U.S. Navy.

President Theodore Roosevelt encouraged the use of wireless aboard Navy vessels, wanting the country to take advantage of any new devices that would help maintain the country's strength. In 1903 the Navy conducted extensive tests at six experimental stations in different coastal locations, and a special naval training school for wireless operators was held at the Brooklyn Navy Yard. The Navy tested not only foreign equipment, but that of Fessenden and de Forest. In the end, the tests were inconclusive, adding confusion to an increasingly complicated situation. The U.S. Weather Bureau was using Fessenden equipment, the Navy used Slaby-

Arco, the Army was using Braun equipment prior to 1903, and Telefunken equipment in 1904. De Forest, with influential navy associations, installed shore stations at San Juan, Puerto Rico, Key West, and Pensacola, Florida; Guantanamo, Cuba, and Colon, Panama.

There was considerable resistance among some old-line navy officers to the introduction of wireless aboard ships, even though tests of equipment during maneuvers had indicated that communication between ships during a naval engagement was a decided tactical advantage. These old-liners felt that wireless would constitute an infringement of their prerogatives to act as commanders of their ships, and that the possiblity of getting orders via wireless could potentially prejudice their command.

After 1902 there was a shift of inventive wireless activity to the United States. The principal reason for this appears to be that competition in the United States for new business was keen and frequently bitter. In contrast, Marconi had a virtual monopoly of wireless communication in the British Empire. In 1904 the British Parliament had passed a law giving the Post Office—which operated all existing communications—ultimate authority over all wireless operations in the United Kingdom. Marconi was the Post Office's delegated supplier and operator of wireless equipment for the British Empire.

There was no such law in the United States; the Navy and Army developed wireless systems of their own, with commercial interests vying for a share in their operation. With no monopoly on wireless equipment in the U.S. both de Forest and Fessenden were able to obtain venture capital to pursue the development of their ideas.

In 1903 de Forest had installed wireless equipment on the yacht of Sir Thomas Lipton, whose *Shamrock III* was attempting to win the America's Cup. Sir Thomas was very much taken with de Forest and his equipment, and invited him to England to demonstrate his apparatus to the British government. Marconi's monopoly was so complete, however, that de Forest had no success in establishing an English operation. But the London *Times* correspondent, Lionel James, en route to the Far East on developments concerning the war imminent between Russia and Japan, met de Forest and became interested in wireless as an aid to war reporting.

On his return to the United States, de Forest dispatched two assistants to the Far East, where a small steamer was equipped with wireless equipment, and from which Captain James reported new events to a land station that had been erected in Shantung.

But de Forest's was not the only company represented. The British in the area were using Marconi equipment, as were the Russians. The Germans were using Telefunken equipment. The Japanese were reported to be using Marconi-type equipment, but claimed it was a Japanese invention. Thus the war of 1904 became the first to be reported on extensively by wireless, and the first in which wireless was used by opposing forces.

The Marconi Company policy of wireless communication only with stations and vessels equipped with Marconi apparatus aroused much antagonism, and as early as 1902 an international incident occurred which led to the request by the German government to have a wireless conference convened.

Prince Henry of Prussia was returning to Germany after a visit to the United States aboard the German vessel *Deutschland,* which was equipped with Slaby-Arco wireless apparatus. Prince Henry attempted to send a message to President Roosevelt as a gesture of courtesy, but the message was not received.

It was alleged that the message was refused because the shore station was using Marconi equipment. Further, there were charges that Marconi stations had deliberately interfered with the *Deutschland*'s attempts to send messages. (W.J. Baker in his *History of the Marconi Company* attributes the failure of the Marconi shore station to pick up the Prince's message to equipment failure aboard the *Deutschland.*) Whatever the reason, the friction that developed between Germany and England was symptomatic of the tensions growing out of the Marconi policy of communicating only with other Marconi-equipped stations.

The Germans therefore convened a wireless conference in Berlin in 1903. The Marconi Company's attempts to monopolize wireless, which were often referred to as Marconism, led to the adoption of a protocol requiring that coast stations handle messages to or from ships at sea without distinction as to the type of equipment being employed. In spite of the objections of Great Britain and Italy, the principle was supported by the seven other nations in attendance, Germany, Russia, the United States, Spain, France, Hungary, and Austria. The 1903 conference had been preparatory, and the delegates returned home to take up the recommendations of the conference with their governments.

A second conference was held in Berlin in 1906. Attended by twenty-nine nations, it accepted the recommendations made by the 1903 delegates, requiring "that all coastal wireless stations should receive from, and transmit to, all shipping regardless of the type of apparatus which the vessels were using" (Baker, *A History of the Marconi Company*).

The 1906 conference adopted a Radio Convention, and annexed a set of radio regulations to this convention. Italy and Great Britain again resisted the concept of mandatory communication between stations, but the conference adopted the resolution overwhelmingly. It was revealed that the British government and the British Post Office had promised in separate agreements to subsidize the Marconi Company for any losses suffered because of the new regulations. The British and Italians attempted to have the system of subsidies adopted internationally, but their efforts failed.

The 1906 Conference Radio Regulations for the first time assigned specific frequencies to certain services. Two frequencies, 500 and 1,000 kHz, were allocated exclusively to the maritime services. Frequencies below 188 kHz were allocated to coastal stations operating long-distance services. The range between 188 and 500 kHz was allocated to military stations, and defined as stations "not open to public correspondence."

Of equal significance, the participants agreed on the term *radio* to

Fig. 23. Hugo Gernsback at the key of a replica he built of "the first home radio ever offered to the American public." (The set, designed by Gernsback and sold by his Electro Importing Company, was sold in 1906; the photograph was taken in 1956.) *Courtesy of Gernsback Publications.*

Fig. 24. This "first home radio advertisement to appear in print anywhere" was in the January 13, 1906, issue of the *Scientific American.*

describe communication through space without the use of wires. The word did not come into general use for many years, and the British continue to say "wireless" to this day.

Among the other agreements of historic significance was the adoption of the distress signal SOS. Contrary to popular belief, the signal does not stand for "Save Our Souls" (or "Save Our Ship"), but was based solely on readability of the signal which, in international Morse code, is ...---...; the Germans had been using SOE, but many of the delegates felt the letter *E*, which is a single dot, was not completely satisfactory. After much haggling, SOS was adopted. The previous distress signal, CQD, continued to be used by some, and during the *Titanic* disaster, both calls were sent. The regulations went into effect July 1, 1908.

The radio amateurs, that group who were interested in wireless communication solely as a hobby without regard to personal financial gain, and who were later to contribute so much to the development of the radio art, had remained fairly unobtrusive during the first years of the new century. Their numbers grew slowly; they were not part of any organized group, and until 1908 there was not any clear line of demarcation between amateur and professional wireless enthusiasts.

By 1908, commercial wireless had become well established. Two-way commercial transatlantic wireless service had been inaugurated by the Marconi Company on October 15, 1907. Bands of frequencies had been allocated by international agreement, and prospects for the future were brightening. In 1908 the first magazine devoted primarily to amateur wireless was published. Hugo Gernsback, a young immigrant from Luxembourg who already was supplying American radio enthusiasts with various kinds of wireless equipment through his Electro Importing Company, began to publish *Modern Electrics,* which would shortly devote itself almost exclusively to the new radio art.

Books by authors such as Victor H. Laughter, Elmer Bucher, and L.W. Bishop helped popularize radio communication. With the help of these media, amateurs began using tuners. This gave them an advantage over many commercial stations, which were often operated with outmoded, inefficient equipment because a handful of organizations held patent rights to the most effective tuners used at that time. The amateurs, many of whom were self-trained, did not need to concern themselves with obtaining rights to use patented apparatus before going on the air.

In 1909 the first national amateur organization, the Wireless Association of America, was formed by Hugo Gernsback through his magazine *Modern Electrics.* Within two years it had nearly 10,000 members, many of them actively on the air. The circulation of *Modern Electrics* jumped from 2,000 to 30,000 by the end of 1910. Other magazines of the day, quick to note the rapidly growing demand, began to carry radio information, including how-to-do-it pieces. The number of manufacturers supplying electrical equipment to amateurs also began to increase, as did the number of radio clubs throughout the country.

Amateur equipment was also becoming more sophisticated. Although the basic apparatus still consisted of the spark transmitter, the components used continued to improve as manufacturing skills increased. Detectors also improved, and by 1909 the coherer had largely been replaced by crystal and electrolytic detectors. In 1911, de Forest's Audion was advertised in *Modern Electrics.*

As their equipment and skills improved, so, too, did the distances covered. In some cases amateur contacts of several hundred miles were recorded, and as transmitter power increased, the strength of the amateurs' signals became greater and greater. It was inevitable that the problem of interference would grow increasingly serious with the passage of time.

There was, during these early days of radio, no legislation limiting the use of frequencies to any particular services. The Berlin Conference of

1906 had allocated some frequencies to specific services, but by and large no such restrictions existed in the United States, and it frequently occurred that an amateur interfered with a station carrying on a commercial service. In the words of Clinton DeSoto, assistant secretary of the American Radio Relay League, in *Two Hundred Meters and Down*,

> If a commercial station wanted to do any work, it was usually necessary to make a polite request of the local amateurs to stand by for a while. If the request was not polite, or if an amateur-commercial feud happened to exist, the amateurs did not stand by and the commercial did not work. Times without number a commercial would call an amateur station and tell him to shut up. Equally as often the reply would be, "Who the hell are you?" or "I've as much right to the air as you have." Selfish? Undoubtedly. And yet, the amateur did have equal right to the air with the commercial, from any legal or moral standpoint. He was seldom interrupting important traffic—contrary to accusations that have been made, there is no authoritative record that amateurs ever seriously interfered with any "SOS" or distress communication; on the contrary, there are instances when the constantly-watchful amateurs heard distress calls which were not picked up by the regular receiving points. And he was even then doing a useful work developing new and better radio equipment through his patronage of the manufacturers of parts and apparatus.

By late 1909 it was obvious that some kind of legislation regulating amateur operations was necessary if commercial and government organizations were to survive the growing wave of interference that was slowly but steadily drowning these services. Amateur stations on the airwaves now outnumbered the navy stations, for example, by about four to one.

Legislation was introduced in Congress in 1909 to limit amateur activities, but was defeated because of the support given amateurs by the American Marconi Company, which argued that American wireless operations were hindered not so much by the amateurs as they were by obsolete equipment. The Marconi Company, in addition to recognizing amateur radio as a vast training ground for operators and technicians, also used the chaotic situation to promote its own equipment.

As a result of the conflicts among the amateurs, Marconi, the U.S. commercial operators, and the government, a major controversy arose, with many newspapers and magazines rallying to one side or the other in the multifaceted dispute. Meanwhile, the number of amateurs was growing steadily, numbering some 10,000 actively on the air. It was evident that so great a number of enthusiasts could not simply be legislated out of existence, and increasingly talk of some kind of compromise began to grow.

Bills were introduced in Congress both in 1910 and 1911 that would have made it illegal for any operator to interfere with existing commercial or government services. Without actually mentioning the amateur specifically, the legislation would have made use of the airwaves by amateurs virtually impossible. For the first time in their history the radio amateurs banded together to oppose the crippling legislation, and their lobbying activities paid off. Bills in both 1910 and 1911 were defeated. But legislation was inevitable.

Technological developments had resulted in the convening in 1912 of another radio conference, held in London. At this conference frequency allocations to various services were made, including radio beacons, weather reports, and time signals, as well as allocations to coast, ship, and government operations.

At the London conference of 1912, the frequencies 500 and 1,000 kHz were retained for use by the maritime services. The new services transmitting radio beacons were allocated any frequency above 2 MHz; two other new services, weather reports and time signals, were allowed to use any frequency below 188 kHz. And 167 kHz was allocated to ships transmitting messages to their home countries where the receiving station was farther away than the nearest coast station.

U.S. legislators, aware of the allocations to be made at the London conference, fitted their amateur legislation to the London proceedings. Rather than legislate the amateurs out of existence, they allocated wavelengths below 200 meters (frequencies above 1.5 MHz) to them, and made it illegal for them to operate above 200 meters without special permission.

This was a subtle way of attempting to destroy amateur radio, because according to the best scientific thought of the day the longer wavelengths were the most useful for radio communication. The shorter they became, the less effective they were thought to be. The consensus was that wavelengths of 200 meters and down were essentially worthless. On May 7, 1912, the Radio Act of 1912 dealt what was thought to be a death blow to amateur radio.

Regulation Fifteen of that act provided that

> No private or commercial station not engaged in the transaction of bona fide commercial business by radio communication or experimentation in connection with the development and manufacture of radio apparatus for com-

mercial purposes shall use a transmitting wavelength exceeding two hundred meters, or a transformer input exceeding one kilowatt; except by special authority of the Secretary of Commerce contained in the license of the station.

Actually, of the many important contributions radio amateurs have made to the radio art over the years, this regulation, the fifteenth, would lead to the greatest contribution of all.

7

A New Radio Era

Before wireless, a ship more than twenty miles beyond the horizon was totally isolated, completely at the mercy of the elements. If it ran into difficulty its only hope for assistance was a chance encounter with another vessel.

Wireless telegraphy was to change all that. At the turn of the century, on January 23, 1900, the icebreaker *Yermak* in the Baltic Sea, instructed by wireless from St. Petersburg, rescued a group of fishermen stranded on floating ice in the Gulf of Finland. The new Marconi Company, restricted by British communications practice from communicating between British land points, equipped many vessels with equipment intended both for safety and commercial use.

In the United States, there was no such limitation on overland communications, though communicating with ships at sea formed the bulk of the new wireless companies' business. In 1903, the Pacific and Continental Wireless Telegraph and Telephone Company had erected stations on Catalina Island and at San Pedro in California, and the de Forest Wireless Telegraph Company of America had stations at Coney Island, New York; Galilee, New Jersey; Cape Hatteras, North Carolina; Key West, Florida; Galveston, Texas; and New Haven, Connecticut. By the end of 1904 stations at Buffalo, New York; Cleveland, Ohio; and Port Huron, Michigan, had been added.

In 1907 the Marconi Wireless Telegraph Company of America had put up stations at South Wellfleet and Siasconset (Nantucket), Massachusetts, and Sagaponack, New York, and entered into intense competition with American-owned companies, chief of which was United Wireless.

After the takeover from de Forest, United Wireless expanded rapidly, and between 1907 and 1912 added twelve stations along the Atlantic and Gulf coasts from Eastport, Maine, to Tampa, Florida, and Port Arthur, Texas, as well as Pacific stations at San Francisco, California; Astoria, Oregon; and Seattle, Washington. The Poulsen (later Federal and still later ITT) company had arc stations in Stockton and Sacramento, California, in 1910, and by 1912 had fourteen stations, the most easterly one in Chicago, and most westerly in Honolulu.

Some of these stations (particularly the inland ones) were used for purely commercial purposes. But most of them were strung around the sea and lake coasts. Intended for ship-to-shore and shore-to-ship communication, they increased the safety of life at sea tremendously.

By 1909, the crew of many a storm-tossed craft had been saved by calling for help by radio. In spite of the numerous sea rescues, however, most owners of ocean-going ships found the cost of wireless too high, and considered it a luxury. Two events within three years resulted in a significant change in the attitude of owners, and in the equipment and personnel that ocean liners carried.

In January, 1909, the 15,000-ton liner *Republic*, with 460 passengers aboard, collided in dense fog some 26 miles southwest of Nantucket, with the Italian vessel *Florida*, bound for the United States with eight hundred immigrants. The *Republic*'s wireless operator, Jack Binns, off-duty at the time, hastened to the wireless room, where he found his equipment partly destroyed by the impact. Switching quickly to an emergency power supply (the main source of power had been wrecked) he began sending the old distress signal, CQD. It was picked up by the United States Coast Guard station at Siasconset and relayed to other ships in the area. Within half an hour, the liner *Baltic*, some 200 miles distant, had received the call and was speeding toward the disaster area. Guided by the bearings Binns was providing, the *Baltic* raced through dense fog, reaching the stricken vessel in twelve hours. By then, the passengers and crew of the rapidly sinking *Republic* had been transferred to the *Florida*.

Had it not been for the wireless, and the courageous performance of Jack Binns, a sea disaster of epic proportions might have occurred. As it was, only five people lost their lives, all as a direct result of the collision. For the first time the world had watched anxiously as a dramatic sea rescue was taking place, participating in the events as they were reported in "extra" editions of the newspapers.

The *Republic* sinking led to the Radio Act of 1910, which made it mandatory for any ship carrying a total of fifty or more people—passengers

Fig. 25. David Sarnoff at his station atop the Wanamaker Building in New York City, where he received the reports on the *Titanic* disaster. *Courtesy of RCA.*

and crew together—to be equipped with wireless equipment and an operator. Similar laws were passed in England. In 1912 the act was amended to include cargo vessels; it required a continuous watch in the radio room and a minimum of two operators on board at all times.

The amendments to the original act were the result of the worst peacetime disaster in maritime history, the sinking of the *Titanic*. In April 1912, the "unsinkable" 46,000-ton luxury liner, on her maiden voyage from England to New York, struck an iceberg while racing through the frigid waters of the North Atlantic, and sank within three hours, with a loss of more than fifteen hundred lives.

The irony of this tragedy was that the sinking occurred practically within sight of another vessel, the *Californian*, which did not pick up the

Titanic's SOS's and CQD's because her wireless operator had turned in for the night. Earlier in the evening, the *Californian* had encountered an ice field and her radio operator had attempted to relay this information to the *Titanic*'s operator. He was told to "shut up" because the *Titanic* was exhanging messages with another station at the time. When the *Titanic* struck the iceberg the *Californian* was within twenty miles of her, and could have been at her side in an hour.

The *Carpathia*, which heard the distress signals almost immediately, was some sixty miles away from the *Titanic* at impact, and did not reach the area until after the *Titanic* had gone down. It was able to pick up 710 survivors from the sea.

One illustrious radio career began on the night of the *Titanic* tragedy. A twenty-one-year-old wireless operator, manning Marconi equipment atop the Wanamaker store in New York, received the first distress signals from the *Titanic*, and promptly made the news available to the world. The young man, David Sarnoff, remained on duty for seventy-two hours, during which time he received the complete list of survivors and other messages from the rescue vessel *Carpathia* and relayed them to an anxious populace. While Sarnoff manned his station, President Taft ordered all other radio stations on the East Coast off the air to prevent interference.

The Radio Act of 1912 heralded the opening of a new era. With a world war looming on the horizon, major changes were taking place in radio science. Work on the triode, which had been at a low level since its invention, was begun anew by a group of scientists and engineers who brought fresh and inventive ideas to its use. Revolutionary new triode circuits were also being pioneered. One of the brilliant and innovative thinkers whose work would materially affect the course of radio development was a radio amateur, Edwin Howard Armstrong.

Born in New York in 1890, by the time he was fifteen Edwin Armstrong was an avid amateur, reading voraciously about wireless and experimenting with new circuits. At nineteen he entered Columbia University, graduating with an electrical engineering degree four years later. At Columbia, Armstrong studied under Michael Pupin, who encouraged his research. It was as a student at Columbia that Armstrong discovered the principle of *feedback*, or regeneration, which would have an important effect on the course of radio for the next several years.

Before the invention of the triode, there was no effective way of amplifying, or making stronger, the radio signal. The crystal or vacuum (Fleming valve) detectors could in no way amplify. Whatever energy was received and turned into sound had to come direct from the transmitting station. The coherer and magnetic detectors, because they used the radio signal to trigger a local source of power, could amplify, but they were so insensitive that crystals or electrolytic detectors were better receivers. The same was true of the magnetic detector. The triode did amplify to a slight extent, which made it a better receiving device than anything used before it.

Armstrong took advantage of the fact that the Audion detector did amplify—that the signal at its output was stronger than the radio wave that came in to the grid. He *fed back* a part of that output to the grid circuit in phase with the incoming signal, so that it helped or strengthened it. The stronger signal in the output fed back more power to the input, with the result that far greater amplification than had ever before been possible was obtained. A few amateurs were able to obtain tubes and made distance records with receivers like that of Figure 26.

Meanwhile, at the Federal Laboratories on the West Coast, de Forest was also trying to amplify the signals received from the various Federal arc stations. In the past, several attempts had been made with mechanical or acoustic apparatus. The line telegraph had long used a *relay*—a magnetic device like the telegraph sounder or the doorbell-like decoherer that supplied the sound in coherer circuitry. Microphones had been used to receive and amplify the weak sound from the telephone receiver. None had

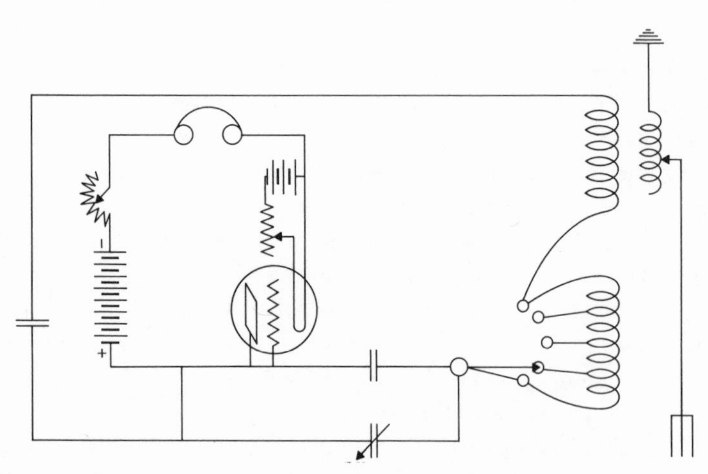

Fig. 26. An early regenerative receiver. *From* QST *Magazine, December, 1916.*

been particularly efficient. De Forest tried to use the amplifying ability of the Audion, hooking up a second Audion after his detector with a type of transformer used on telephone lines. Eventually he found that a higher vacuum would allow the tubes to continue working with as much as 120 volts applied. The higher voltage produced enough of an increase in amplification to make a practical amplifier of the Audion. It was these "hard" tubes that de Forest demonstrated to AT&T when he attempted to sell the wire rights to the telephone company. De Forest and his assistant Van Etten discovered that the output of the Audion could be turned back on its input, producing the results that Armstrong was producing in the East.

There was a long and intense patent struggle between de Forest and Armstrong. It drew so much attention that the two emerged as the only possible inventors of regeneration. It is, however, probable that several persons made the discovery independently at about the same time. Meissner in Germany and Round in England both claimed independent discovery, as did a half dozen persons in the United States and at least one in Canada. From a piece of testimony in a court case, it would appear that the American inventor John Hays Hammond had produced the regenerative effect a year or so before either Armstrong or de Forest thought of it.

Meanwhile, de Forest was selling audio amplifiers with his new improved tubes, as was Federal, which had "shop rights" to de Forest's invention since he had made it while working for the company.

The patent battle, based on the question of who was first, lasted for years and through more than one judicial decision, forcing one inventor and then the other to stop using the technique. At last, in 1934, the Supreme Court upheld de Forest. By that time Armstrong had acknowledged that de Forest had been the earlier discoverer of the principle, and based his action on the contention that de Forest had obtained only audio (sound) oscillations, and did not understand that the Audion could oscillate at radio frequencies. This, in the words of Supreme Court Justice Cardozo, who decided the case: "De Forest denies. He maintains . . . that upon discovering the effect of feedback in generating sustained oscillations of the plate, he understood at once that by controlling the inductance or capacity in the oscillating circuit he could also control the frequency. . . . There is evidence that in August, 1912, he discussed with his assistants the possibility of using sustained oscillations of the Audion in generating and transmitting radio waves as well as those of audio frequency . . . and particularly that on April 17 (of 1913) . . . he received a clear note, the true heterodyne beat note, from the radio station at San Francisco Beach with the aid of the coupled circuits. The entry in his notebook the same day tells us, 'This day I got the long looked for beat note.' This was long before he heard of Armstrong or of like experiments by anyone." The opinion, which was the final one in the litigation, was dated May 21, 1934.

Many of the prominent engineers of the time, however, believed it was Armstrong who had conceived and fully appreciated the principle of feedback and who deserved the credit. He had been awarded the Medal of Honor of the Institute of Radio Engineers in 1918, chiefly because of his discovery of regeneration, and after the Supreme Court Decision of 1934, returned it to the IRE. The Institute reconfirmed the award, and sent the medal back to Armstrong.

Regeneration was not the only subject of patent fights. The American Marconi Company brought an action against de Forest in 1914, claiming that the Audion infringed the Fleming valve patent, owned by Marconi. The Audion, the complainant stated, was not an invention, but simply a modification of the Fleming valve, and the patent was therefore invalid. In 1916, after a long struggle, the court decided that both the Fleming valve and the de Forest Audion were inventions, upheld both patents, and ruled that in making Audions both de Forest and Marconi infringed each others' patents. The result threw the industry into hopeless confusion until the Fleming valve patent ran out in 1922. The net practical result was that neither de Forest nor Marconi could make Audions. This led to some absurdities. In 1919 the Marconi Company licensed a San Francisco company to make Fleming valves. De Forest then ordered Audions from the company (Moorhead) and sold them to American Marconi!

These were the first vacuum tubes to reach the American public. In the Gernsback publication, *Radio Amateur News,* of August 1919 and for several months thereafter, the *Marconi VT* was advertised at seven dollars each. They were offered by "Marconi Wireless Telegraph Co. of America (Sole distributors for de Forest Radio Telegraph and Telephone Co.)."

Shortly afterward, Moorhead advertised detector and amplifier tubes, and in early 1920, Audiotron offered double-ended, double-filament Audions. In December 1920, in the same magazine (now *Radio News*) Audiotron advertised C300 and C301 tubes at five dollars and six dollars each, and in March 1921, Cunningham offered the tubes under its own name, adding the C302 (5 watts) at eight dollars and the C303 (50 watts) at thirty dollars.

8

Progress and Problems

The sinking of the *Titanic* spurred both radio research and recognition. Legislation requiring ocean-going liners to carry wireless equipment was enacted in many countries. Those that already had such legislation increased the requirements, making wireless mandatory on smaller vessels and assuring a twenty-four-hour watch on large passenger-carrying vessels. (Loss of life on the *Titanic* would not have been as great if the *Californian* had been furnished with more than one operator.) The outbreak of World War I, two years later, further accelerated research and development. De Forest had learned, in California, the advantages of pumping his tubes to a higher vacuum. The discovery was made on the East Coast independently by Harold D. Arnold of AT&T and Langmuir of General Electric. Arnold further replaced the tantalum filament of the original Audion with an element coated with metal oxides, which emitted more electrons than the older type.

Meanwhile, other scientists were making contributions to vacuum-tube technology. The tungsten filament was developed by William D. Coolidge of General Electric, and many laboratories conducted research on interelectrode physics. Simultaneous developments by General Electric and AT&T led to many patent-infringement conflicts during the decade following the *Titanic* sinking. Many of these were settled out of court with cross-licensing agreements which gave contesting companies rights to each others' processes.

In 1913 de Forest attempted to sell AT&T the rights to the use of the Audion as a telephone repeater. He demonstrated his Audion, but heard nothing from the company for ten months. Apparently AT&T was not interested.

Then he was called on by an attorney, Sidney Meyers, who told de Forest he represented clients interested in purchasing rights to the Audion. De Forest, with bankruptcy and the possible forced sale of the Audion patent together with the rest of the assets of the company staring him in the face, was willing to listen. Meyers offered him $50,000 for the rights, on a take-it-or-leave-it basis, and de Forest recommended that the offer be accepted. Special stockholders meetings of the two de Forest concerns, the Radio Telephone Company and the North American Wireless Telegraph Company, approved the transfer, and the deal was made.

Then, de Forest says, "the truth leaked out. Meyers turned out to be none other than the American Telephone and Telegraph Co.! I later learned, on reliable but unofficial authority, that the directors of AT&T had voted to pay as high as $500,000 for the wire rights under the Audion patents."

While there is little doubt that de Forest was overreached, the deal was not as bad as some have made it out to be. He had not sold the patent—merely the right to use it as an audio amplifier on telephone lines. Sales of further rights gave him, in 1914, $90,000, and in 1916, $250,000, and he still retained some rights in the Audion.

It was at about this time that Irving Langmuir of General Electric and A.T. Arnold of AT&T were experimenting with Audions in which the gases had been more completely evacuated. (Arnold's work may have started in the ten months de Forest was waiting for an answer to his offer of sale to the telephone company.) Langmuir, inventor of a greatly improved vacuum pump, evacuated tubes to a point that produced a "pure electron discharge" and together with E.F.W. Alexanderson, who had suggested the experiments, came up with an improved electron tube, which they called an "electronic amplifier" (as distinguished from Alexanderson's magnetic amplifier, which he used as a modulator in arc telephone transmissions). They attempted to patent this as an invention, but the courts held that an improvement of the vacuum tube was not in itself invention. But the new "electronic" concept and the use of "hard" tubes by AT&T marked the end of the "ionic" concept which had probably hampered de Forest in his work, and led to the revolution that vacuum tubes would cause.

Meanwhile, one of de Forest's earliest companies, reorganized as the United Wireless Company after tossing de Forest and his Audion patent out, had fallen on bad days. Stock jobbing, the occupational disease of communications company officials, had put United Wireless in a precarious financial position. The president and several officers were convicted of selling stock under false pretenses, the company went into bankruptcy, and was later taken over by American Marconi Company in an action that made them a power in the American wireless field for the first time. The number of their land-based stations increased from 5 to 50, and shipboard stations multiplied from a handful to about 400.

The takeover of United Wireless by American Marconi led—more or less directly—to a scandal in Britain which proved that difficulties involving sales of communications company stocks were not an American monopoly. In 1912 a preliminary agreement was made between Marconi Company and the British government to construct a round-the-world series of wireless stations that would link the parts of the British Empire. During the short period between the late summer of 1911 and the spring of 1912, Marconi shares more than tripled in value. Charges of collusion between Sir Rufus Isaacs, British Attorney General, his brother Godfrey Isaacs, Managing Director of the Marconi Company, and Sir Herbert Samuel, the Postmaster-General (the Post Office has charge of all British communications) were made by members of the press. Questions were asked in Parliament, and all queried denied having any interest, direct or indirect, in Marconi stocks.

Meanwhile, American Marconi, which was to some extent controlled by the British company, had decided to increase its capital about $6 million, to raise the cash to buy United Wireless and to improve its facilities. The money was to be raised in England, with Marconi and Godfrey Isaacs chiefly responsible for placing the shares. Among those who bought, directly or indirectly, the shares sold by Isaacs were his brother, the Attorney General; David Lloyd George, the Chancellor of the Exchequer; and Lord Murray, a leader of the ruling Liberal Party.

The Parliament members who asked questions about the involvement of high political figures in Marconi transactions knew nothing of their dealings in American Marconi shares. But the secret leaked out and resulted in what became notorious as the *Marconi Scandal.* The individuals involved insisted that they had understood that the questions asked in Parliament were about the British company only, and they had seen no reason to mention American Marconi.

Nevertheless Parliament took the accusations very seriously and appointed a select committee to investigate the situation. The investigation resulted only in worse confusion. The committee, controlled by the majority Liberal Party to which Isaacs, Lloyd George, and the others under inquiry belonged, reported that none of the principals profited in any way through the stock transactions. A minority report, which differed greatly from the official findings of the committee, was issued, and strangely enough, that report was the one on which discussion was usually based when the affair was debated in Parliament. In it was the statement: ". . . the impropriety of which the Ministers . . . were primarily guilty was that of making an advantageous purchase of shares from a government contractor."

Another committee, appointed to investigate the engineering angles of the Marconi-government agreement, made the naïve determination that the Marconi Company was technically better qualified to provide the service envisioned in the contract for a worldwide ring of wireless stations than any of the rival systems then operating.

This decision was not entirely due to the British conviction that anything foreign must necessarily be inferior. Some of the "technical" reasons were political. It was considered "unwise" to put construction of the Empire's communications system in the hands of a foreign company, such as Germany's Telefunken. Germany's intense naval program was one of the reasons Britain was trying to improve its global communications system—the coming of World War I was already beginning to cast its shadow ahead. The Poulsen arc, though supported by the acknowledged expert A.A. Campbell Swinton, had yet to prove itself, and while de Forest's spark system was more modern than that of Marconi, the British company insisted that de Forest was infringing Marconi patents.

Marconi himself, though absolutely untouched by the scandal, was deeply distressed over the aspersions that had been cast at his company. Construction of the series of wireless stations was finally approved by Parliament, but World War I erupted before a great deal had been done, and it was ten years before Marconi's dream of a network of wireless stations linking all the British dominions would be realized.

By 1915, AT&T had used the vacuum-tube repeater to carry voice across the United States and had experimented successfully with a 550-triode vacuum-tube transmitter that sent voice signals over 3,000 miles, to Paris and Hawaii. But major problems with atmospherics led one company official to conclude that radio telephony would be useful only to reach areas where no cable facilities existed.

The General Electric Company, too, had been experimenting with long-distance wireless telephony. While AT&T limited its research with tube transmitters, GE went ahead with large transmitting triodes, as well as the alternator.

Fig. 27. Opening of the transcontinental telephone, between New York and San Francisco, September 29, 1915. Seated is Theodore N. Vail, President of AT&T Company. *Courtesy of Bell Telephone Laboratories.*

Alexanderson had further improved the alternator, developed a magnetic amplifier and by 1916 had an improved antenna with multiple-tuned circuits which tremendously increased the efficiency of the transmitter.

Because of Alexanderson's pioneering efforts, GE had by 1916 a complete and highly efficient transmitting and receiving system. These were demonstrated to Marconi, who was so taken by the alternator that he offered to negotiate with GE for the exclusive rights to it. The outbreak of war led to the postponement of negotiations, but by 1917 GE had built and installed for Marconi a 50-kW alternator at the New Brunswick, New Jersey, transmitting plant of the American Marconi Company. Shortly after, GE installed a 200-kW alternator at the same site, with the call WII, which became famous as one of the most powerful stations in the world.

DEVELOPMENTS DURING THE WAR

One of the first acts of the belligerents was to cut the undersea cables that linked the various continents of the world. The almost immediate result was to increase the reliance on radio for long-distance communication. The high-powered spark transmitter at Nauen, near Berlin, became Germany's primary radio link with the outside world. The British

Admiralty took control of Marconi-operated stations and immediately began using them to carry military traffic. The great increase in the volume of radio communication during the war greatly increased the demand for personnel. Here the relatively new hobby of amateur radio paid its first dividends. Literally hundreds of experienced operators were waiting in the wings to be called to duty. They knew the code, used it, and many were self-trained in electronics. They fitted immediately into the gap created by the increased demand for radio operators; they manned equipment which was rapidly being installed in airships, at the battlefront, and later in aircraft.

Direction-finding (DF) equipment played an important role during World War I. In 1912, Marconi had purchased the patents for a direction-finding method from two Italian scientists, E. Bellini and A. Tosi. They used two fixed coils with a rotating search coil to ascertain the direction from which signals were coming. These early direction-finding receivers, accurate to within half a degree, played an important part in one of the great battles of World War I.

By 1915 direction-finding and content-monitoring stations ringed the German Empire, maintaining a constant surveillance of radio activities. The Germans' output was checked, and reported to British Intelligence. These radio networks maintained a close watch on German undersea, surface, and air operations. The Germans, who had their own much more primitive DF equipment, communicated freely among themselves by radio, and German naval vessels used low-power radio telegraph communication without any inhibitions whatsoever when they were close to home port.

The German navy, whose strength was considerably below that of the Allies, kept fairly close to its home bases, striking quickly close to home with short forays into the North Sea. In May of 1916 British DF stations observed a sharp increase in traffic from a German warship anchored at Wilhelmshaven, one of Germany's chief naval stations. Although British Intelligence could not translate the traffic, which was coded, it reasoned that the increased volume constituted sailing orders to other warships. This supposition was reinforced by a change in position of the transmitting vessel later the same day. It had moved toward the open sea. Accordingly, the British fleet was ordered toward the port of Wilhelmshaven with orders to intercept the expected flotilla of German warships.

The following day the battle of Jutland was fought. Although the battle was by no means a British victory, the Germans' loss of one battleship, five cruisers, and five destroyers changed the course of naval warfare for the duration, and after Jutland, the German navy seldom ventured out of its home ports.

By 1916 radio was being used by reconnaissance aircraft reporting to ground-based infantry and artillery units. The first paratroops were used in 1916. Paratroopers were dropped behind German lines with radio sets strapped to their backs. Their mission was to report German troop movements and other intelligence information to their home units. The first dirigibles to be equipped with radio equipment also carried homing pigeons. The past and future were thus linked on the airships, which themselves were like links in a chain.

When the United States entered the war, the reservoir of radio amateurs also proved a godsend to the military, and the great pool of devoted and eager amateurs who volunteered their services in droves to satisfy the country's almost insatiable appetite for trained technicians were an important factor in the success of the U.S. Signal Corps.

The Radio Act of 1912 empowered the President to shut down any radio station if its operation were not in the best interests of the country. In April 1917, with the entry of the U.S. into World War I, all licensed amateur stations in the country were ordered off the air. The President also ordered the Navy to take over the operation of all commercial stations not already controlled by the Army.

Compliance was immediate and complete. All equipment was turned over to the military; officials of the commercial companies and the leaders of the several budding amateur organizations made available to the government all technical talent, including operators and research workers. The military were faced almost immediately with a critical shortage of radio officers, instructors, and men capable of operating radio equipment. The commercial companies had limited personnel available, but the large reservoir of highly skilled amateurs, trained and ready to operate, quickly moved in to fill the gap. Within a matter of weeks, five hundred operators were recruited through the American Radio Relay League. The League was joined in its recruitment efforts by all radio magazines, and the effort to keep trained radio technicians supplied to the armed forces was a great success.

The most significant American technical contributions to the development of the radio art during the war years were made by Captain Edwin H. Armstrong, who had already made a major contribution with the regenerative or feedback principle. Armstrong was serving in Paris as chief of the

Signal Corps Radio Laboratory. While working on a method of receiving and decoding German radio messages, Armstrong and a Frenchman, Lucien Levy, discovered a method of reception, the *superheterodyne*, which was to revolutionize radio receiver technology.

The superheterodyne (supersonic heterodyne) principle combines two frequencies—one of which is the incoming signal and the other is one generated by a local oscillator—in a circuit to produce a third frequency, the sum or difference of the two. Fessenden had introduced heterodyning, but—partly because there was no simple, light, and cheap equipment to produce the heterodyne signal—nothing much was ever done about it.

But by 1917 some vacuum tubes were being used as oscillators to produce continuous-wave radio frequencies, and the heterodyne principle was combined with the oscillator to develop the superheterodyne circuit. An incoming signal is mixed with the output of a vacuum-tube oscillator in the receiving circuit to produce a lower intermediate frequency (i.f.). The i.f. is then amplified, converted to an audio frequency, amplified again, and fed into a speaker.

The exact circumstances of the superheterodyne's invention are somewhat obscure. In the English-speaking world Edwin Armstrong is the sole inventor of the circuit; in areas where French culture prevails, the inventor is Lucien Levy. The matter is further complicated by the fact that they worked in Paris at the same time, corresponded with each other, and met personally. Was the superheterodyne a joint invention? Did one of the two copy the other's work? Or were they working independently and comparing notes from time to time, as Fessenden and Alexanderson did during the development of the alternator? For those interested in firsts, the Armstrong patent, No. 1,342,885, was applied for February 8, 1919, and issued June 8, 1920, to "Edwin H. Armstrong, a citizen of the United States, now residing in Paris." Brevet d'Invention 506,297 was applied for October 1, 1918, and issued to Lucien Levy May 27, 1920.

The superheterodyne principle made the major portion of the amplification process independent of the frequency of the original signal, because most of the amplification was designed to take place at the i.f. By varying the frequency of the oscillator tube, a fixed i.f. could be generated independent of the signal frequency. The great advantage of the circuit was that the signal frequency was modified by the oscillator to fit the amplifier, rather than modifying the amplifier to fit the incoming signal. Amplification could take place at a frequency where the equipment was most efficient, and more selective circuits could be used if it was not necessary to retune them for each change of signal frequency.

Although Armstrong patented his circuit in 1920, it was not widely adopted until some seven or eight years later, with the development of more versatile vacuum tubes, and the use of a higher intermediate frequency than was considered practical in the earlier superheterodynes. The circuit as Armstrong first designed it required eight to ten tubes. With the development of the tetrode several years later, fewer tubes were needed.

9

The Birth of RCA

The takeover of all commercial stations during the war led indirectly to one of the major developments in American communications history—the formation of the Radio Corporation of America. One of the most important stations taken over by the Navy was the American Marconi Company installation, WII, in New Brunswick, New Jersey. The station had been outfitted with a 50-kW Alexanderson alternator, and eventually put a 200-kW alternator into operation.

The alternator was becoming the leading form of long-distance wireless communication, more reliable than the arc or spark. The New Brunswick station had become famous throughout the United States, parts of Latin America, Europe, and Africa. Battleships in many parts of the world could hear its signal, which was so powerful that even some of the portable sets used in the fields of France could pick up its transmissions. As part of its schedule the station, whose wartime call letters were NFF, transmitted daily news broadcasts by radiotelegraph.

In January 1918, the radio medium stepped into a new and revolutionary role: President Wilson's speech to a joint session of Congress, in which his Fourteen Points (which formed the basis for ending World War I) were first enunciated, was carried by NFF. Powerful stations at Darien, Panama, and San Diego also carried the message. Thus, for the first time, a message of historic importance was carried by radio, girdling the globe in one hour and forty-five minutes, the time it took to transmit the 2,700-word address by radiotelegraph.

Wilson's Fourteen Points captured the imagination of much of the world, which saw in them a way out of the bloodiest conflict in human history. The American President was admired and respected, and with the rapid development of radio his views could be communicated to all corners of the Earth with unprecedented speed. The 200-kW Alexanderson alternator had gone into service at New Brunswick in 1918, and Wilson continued speaking to the world through this transmitter. Many of his words were directed toward the German people themselves, and ultimately their impact contributed to the fall of the German Empire.

In late 1918 Wilson sailed for Europe on the liner *George Washington* to attend the peace conference. The ship was equipped with the most modern radio equipment. The *George Washington* was escorted by the battleship *Pennsylvania*, which had the most powerful transmitter afloat. Radio telephone contact was maintained constantly between the two ships, and Wilson's messages were beamed to receiving stations in the United States. The President was thus able to maintain continuous communication with the United States, the first time such a feat was possible for a national leader.

Wilson was very much impressed by wireless. The role it had played in bringing his words to the world, the impact it had on the German nation, the fact that he had been able to maintain continuous contact with the United States while he was on the high seas, led him to the conclusion that it was in the national interest to maintain a position of preeminence in radio communication. His beliefs set off a surge of behind-the-scenes activity to safeguard the postwar radio position of the United States.

Before the United States entered the war, the British-controlled American Marconi Company had been conducting negotiations with General Electric to purchase a substantial number of Alexanderson alternators. As part of the agreement, GE was to grant Marconi exclusive use of the alternator, thus blocking its use by other organizations. This would have given the Marconi interests a dominant position in international radio communication. Since foreign interests (chiefly British) already dominated international cable communications, it was felt that turning radio control over to a foreign organization would be harmful to the United States.

Accordingly, a series of high-level meetings was held to discuss methods of blocking the sale of the Alexanderson alternators to the British. Matters were brought to a head after World War I, when measures which would have brought the ownership of all domestic wireless stations under the control of the U.S. government were proposed. Strong opposition to this legislation developed among amateurs in the United States, who had waited eighteen months to reactivate their equipment. The strongest arguments against government control of radio lay in the fact that it gave little

Fig. 28. The Alexanderson alternator, the most advanced transmitter of the period of World War I. *Courtesy of RCA.*

opportunity for independent research on the part of the youth of the nation. It was pointed out that Marconi himself was just a boy when he first developed his wireless and that all the major contributions to the art since then had been made by private individuals rather than by government employees.

The chief speakers at the committee hearings on radio legislation were Edward Nally, vice-president of the American Marconi Company, who argued that the legislation would set the government up as a monopoly in a commercial business, and Hiram Percy Maxim of the amateurs' association, the American Radio Relay League, who staunchly supported the rights of the individual to independent research and activity. Early in 1919 the legislation was tabled by the committee and by the end of 1919 radio amateurs were back on the air.

The defeat of any legislation that would give the government control of radio left the door open to the Marconi Company to take control of the alternator, for although the government had not yet relinquished control over domestic stations, it was inevitable that it would do so very shortly. The war had suspended negotiations between GE and Marconi, but now talks resumed. Marconi wanted exclusive rights to the alternator; GE offered to provide them on a royalty basis. Then Marconi offered a compromise—outright purchase.

In a letter to Franklin D. Roosevelt, then Acting Secretary of the Navy, Owen D. Young of General Electric Company advised the Navy of the status of negotiations. In April 1919 Roosevelt asked Mr. Young to attend a meeting in Washington at which the proposed sale would be discussed. The meeting, held in early April 1919, was attended by Mr. Young; Charles Coffin, Chairman of the Board of GE; Commander Sanford Hooper, head of the radio section of the Navy's Bureau of Engineering; and Admiral William Bullard, Director of Naval Communications.

Admiral Bullard had been in Europe with President Wilson, attending the peace conference of 1919, and described Wilson's views: the President had been particularly impressed by the fact that news from the United States was almost instantaneously available all over Europe; radio crossed international frontiers as if they didn't exist, and made world communication without censorship or restriction an instant reality. It was Wilson's thesis that power on an international scale would depend on influence over transportation, petroleum, and communication. Wilson felt that the British navy assured that country's domination of transportation, while U.S. oil development assured American domination of world petroleum. If the U.S. allowed foreigners to gain control of radio communication, this country would be second to the British, having fallen behind in two of the three major categories of influence.

Admiral Bullard pointed out that if the alternator was added to the already imposing facilities of the British it would guarantee Great Britain a virtual monopoly over world communication. The General Electric Company agreed to this premise, and maintained that it would much prefer dealing with an American corporation. But the fact remained that there was no corporation other than the American Marconi Company that was in a position to purchase and operate such costly equipment as the alternator. The GE Company was, after all, in business to sell equipment on behalf of its shareholders. It had spent millions of dollars developing the alternator, and would now be left holding the bag. The Chairman of the GE Board, Charles A. Coffin, agreed that it would not be in the best interests of the U.S. to turn control of the alternator over to a foreign country. But what could he do? What alternative was there?

Admiral Bullard then came forth with a stunning proposal: that GE itself take over and lead in the formation of a company whose primary function would be world communication. Such a company would become a major customer of GE, which was engaged solely in manufacturing equipment and was itself not interested in communication. The proposal was exciting and fired the imagination of Owen Young. But the problems were manifold.

First, the government was still in control of radio communication, and the possibility still existed that legislation might be enacted to place all radio communication permanently under the control of a federal agency. Such legislation was supported by Secretary of the Navy Josephus Daniels. The possibility was not considered seriously because of President Wilson's support of the meeting between General Electric and navy officials. If the President backed Daniels he would not have taken that position. The conferees concluded, therefore, that Wilson would throw his support behind the formation of an American communications company.

Second, American Marconi had had a virtual monopoly over wireless communication in the United States before the U.S. government takeover, and indications were that if the seizure were lifted, Marconi would once again be in a very strong position. On the other hand, it was apparent that if top United States officials, including the President, had agreed that a foreign power should not control domestic radio communications, it would be difficult, if not impossible, for the Marconi Company to do so.

It was also becoming apparent, in 1919, that the spark transmitter, which Marconi operations relied upon almost exclusively, was fast becoming obsolete. Both the Poulsen arc and the alternator were superior to the spark, and since U.S. companies had rights to the arc and control of the alternator, the Marconi operation would soon be at a decided disadvantage.

Third, the patent litigation situation was confused; no one company held rights to a large enough number of patents to form a completely modern system. Vacuum-tube litigation now involved not only Fleming and de Forest, but Langmuir and Arnold as well; Armstrong and de Forest were contesting feedback. GE owned the rights to the alternator, but Westinghouse had the rights to the heterodyne circuitry of Fessenden, and AT&T had purchased some of the rights to de Forest's Audion. It appeared that unless a single company had access to the radio patents necessary to build a complete system, confusion in American radio communication would reign for years to come.

It was agreed, therefore, to form an all-American company whose principal business would be radio communication. Because the U.S. government still controlled all radio patents under the still-in-effect wartime powers, and since under the same powers it still controlled commercial radio, it was stipulated that the government should be a partner of the infant company. It was recommended that a member of the U.S. Navy should sit on its board of directors.

Admiral Bullard and Acting Secretary of the Navy Roosevelt supported the arrangement, but when Secretary of the Navy Josephus Daniels returned from Europe, he opposed it on the grounds that a gigantic monopoly would be the result. This argument was countered with the fact that Britain already had a cable monopoly and also threatened to dominate radio in the same way, unless she were stopped. Daniels then changed his position; if his efforts to place radio under the control of the government failed he would support the formation of an American communications corporation.

Officials of GE then consulted powerful Senator Henry Cabot Lodge, who enthusiastically supported the plan. Without waiting for official government sanction to do so, GE went ahead with plans for the formation of a giant American corporation. GE officials approached the American Marconi Company and offered to buy out their assets completely. The Marconi Company had been keeping abreast of the situation as it was developing, and it was evident that the scheme had the backing of the U.S. government. Marconi therefore agreed to sell its assets. Since no one in GE had prior experience in radio communication, it was logical to offer the top executives of the American Marconi Company positions in the new corporation. By late summer 1919, the British Marconi Company had agreed to sell its holdings in the American Marconi Company to General Electric, and by fall 1919, the Radio Corporation of America (RCA) was granted a charter in the state of Delaware.

Owen Young, chairman of the board of the new corporation, immediately launched a program stabilizing the patent situation. By mid-1920 Young had reached a cross-licensing agreement with AT&T, whereby all present and future radio patents of each company would be available to the other for a period of ten years on a no-royalty basis.

Westinghouse, meanwhile, had been attempting to set up a competitive system of its own, and had purchased the feedback (regeneration) and superheterodyne patents of Armstrong, even though he was still in litigation with de Forest over the feedback patent. Westinghouse also controlled the patents on most Fessenden inventions.

But Young, as head of RCA, had moved quickly, concluding agreements with British Marconi and other foreign communications carriers. This made the task of establishing a rival system by Westinghouse very difficult. Rather than engage in mutually harmful cutthroat competition, the two companies joined forces and came to an agreement. Part of this agreement required that RCA purchase 40 percent of its radio equipment from Westinghouse and 60 percent from GE.

Another agreement was quickly concluded with the United Fruit Company, which also held some patents on radio equipment. By 1921, RCA controlled some two thousand patents, including most of the important radio patents of the day. One of the executives of the American Marconi Company became commercial manager of the infant Radio Corporation of America. His name was David Sarnoff.

10

Broadcasting Begins

In 1916, David Sarnoff, the assistant traffic manager of the American Marconi Company, had written a memorandum to his general manager that has become famous in the annals of radio broadcasting. It began, "I have in mind a plan of development which would make radio a household utility in the same sense as a piano or phonograph. The idea is to bring music into the home by wireless."

Sarnoff described the "radio music box" as an instrument that could be used to listen to music, lectures, and firsthand accounts of world events as they were happening. The memorandum concluded with a remarkable prediction: "Should this plan materialize it would seem reasonable to expect sales of one million 'radio music boxes' within a period of three years. Roughly estimating the selling price at $75 per set, $75 million can be expected."

The Radio Corporation of America, with David Sarnoff as general manager, began to manufacture radio receiving sets for the home in 1922. During the first three years of operation, sales of home receivers amounted to $83 million.

THE "FIRST BROADCAST STATION"

But it was a radio amateur, Dr. Frank Conrad, and a visionary executive, H.P. Davis, both of the Westinghouse Electric and Manufacturing Company, who brought the dream of broadcasting to reality. During the war, Westinghouse had undertaken extensive research into the development of radio telephone equipment utilizing vacuum tubes. Dr. Conrad was placed in charge of the program. He had become interested in radio in 1915 as a result of a wager with a friend about the accuracy of his watch. To check its accuracy, Conrad built a receiver so that he could hear the time signals from the naval observatory at Arlington, Virginia.

His new hobby fascinated him, and he constructed a small telephone transmitter, which operated from his home at Wilkinsburg, Pennsylvania. Conrad was licensed by the Department of Commerce on August 1, 1916, and given the experimental call letters 8XK. The station was shut down by Dr. Conrad in April 1917, when the United States entered the war. It operated sporadically during the war under special military authorization because Westinghouse was manufacturing and testing military equipment, including wireless sets. In October 1919 the ban on amateurs was lifted and Dr. Conrad again obtained his call, 8XK.

To a great extent the conversation of radio amateurs consisted (as it does today) of exchanges of information about the strength of the signals received and the type of equipment being used. Although he found such data interesting technically, Dr. Conrad soon became bored with the routine. As part of his radiotelephone experiments he placed his microphone before a phonograph and began playing records over the air.

Fig. 29. The transmitting equipment in Dr. Frank Conrad's garage in Wilkinsburg, Pennsylvania. This is the original home of radio station 8XK. *Courtesy of Westinghouse Broadcasting Co., Inc.*

Conrad was deluged with mail. Amateurs from around the country wrote telling of their pleasure, and requesting that he play specific records. Within a short time after the inauguration of his music program, requests became so heavy that Dr. Conrad had to change his format. Instead of honoring individual requests he would "broadcast" records for two-hour periods every Wednesday and Saturday evening. Dr. Conrad borrowed the word from the gardening term, which means to cast or scatter over a wide area, as in sowing seed, and used it in his programs.

Before long, Dr. Conrad had depleted his supply of records, and an enterprising local entrepreneur offered to make a steady supply of records available to him if he mentioned the store on the air. Conrad agreed, thereby establishing the first instance of advertising by radio. By September 1920 interest in the broadcasts had become so widespread that a Pittsburgh department store, the Joseph Horne Company, advertised in the Pittsburgh *Sun,* calling attention to Dr. Conrad's programs and revealing that the store was selling radio receivers (crystal sets) that could tune in the broadcasts.

Harry P. Davis, vice-president of Westinghouse, had been following the activities of Dr. Conrad with interest, and, reading the Horne ad, realized that radio broadcasting was potentially the most remarkable means of collective mass communication in history. For the first time, vast numbers of people could be reached simultaneously and instantaneously. All that stood in the way was the lack of receivers and programs. The Horne Company ad had run September 29. By October 16, Westinghouse had applied for a license to broadcast from Pittsburgh, and on October 27, the Department of Commerce authorized the station to operate and assigned the call letters KDKA. Thus began two weeks of feverish activity, for it was decided that the inaugural program would carry the November presidential election returns.

Arrangements were made with the *Pittsburgh Post* to obtain the returns and phone them to the station. To publicize the event, a loudspeaker and a receiver were installed in the ballroom of a suburban Pittsburgh community center.

The KDKA transmitter consisted of two 50-watt oscillator tubes and four 50-watt modulators. The station operated on a wavelength of 500 meters (600 kHz).

Dr. Conrad, whose work had inspired the project, was not present at the historic first broadcast. Fearful that the quickly assembled KDKA transmitter might break down, he stood by at his home, prepared to man

Fig. 30. The Harding-Cox election returns broadcast, November 2, 1920. *Courtesy of Radio KDKA.*

his amateur station 8XK, if an emergency should occur. Fortunately, the equipment functioned perfectly, and the broadcast continued from 6 P.M., November 2, 1920, until noon the following day, hours after the Democratic candidate, James W. Cox, had conceded to Senator Warren G. Harding.

The broadcast caused a sensation throughout the country, and newspapers proclaimed the history-making event. KDKA was immediately hailed as the nation's first broadcast station, a claim that was challenged almost immediately by a number of other contenders. Some of the claims were almost as technical and artificial as those that cropped up in the numerous patent suits to which the industry had become accustomed, centering around regularity of broadcasts, continuity of operation, etc.

Stations claiming priority were WWJ of Detroit, and 9XN (now WHA) of Madison, Wisconsin. It was operated by Carl Jansky and Professor Carl M. Terry at the University of Wisconsin, in 1919.

Later it appeared that de Forest had carried on regular broadcasts from his High Bridge station in New York City in 1916. He had actually beaten Dr. Conrad to the punch and had announced the 1916 election returns. Dr. de Forest says that, "I, as chief announcer, proclaimed at 11 o'clock, just before I closed the station, the election of Charles Evans Hughes!"

But a contender from the Pacific Coast, Charles Herrold, was already transmitting regular programs for years before even de Forest came on the scene. Attempting first, in 1909, to use spark telephony, he found it necessary to use an arc transmitter to be understood.

Thereafter he ran regular half-hour daily broadcasts. (The half-hour limit was set by the microphone, which became too hot to work after a half hour.) Finally in 1921, the government got around to issuing licenses, and gave Herrold the call KQY, and a frequency of 833-kHz (360 meters). Herrold had to shut down; the arc could not be made to work on so high a frequency. Getting back with a 50-watt tube station, he continued until 1925, then sold the station, which is now KCBS.

Thus KDKA was in the same situation as Marconi, given credit for something it had never claimed. But also like Marconi, the backers of the station could point out that when KDKA started, broadcasting started! Within a year three other Westinghouse stations were on the air, and the

Fig. 31. The original KDKA broadcast transmitter. *Courtesy of Westinghouse Broadcasting Co., Inc.*

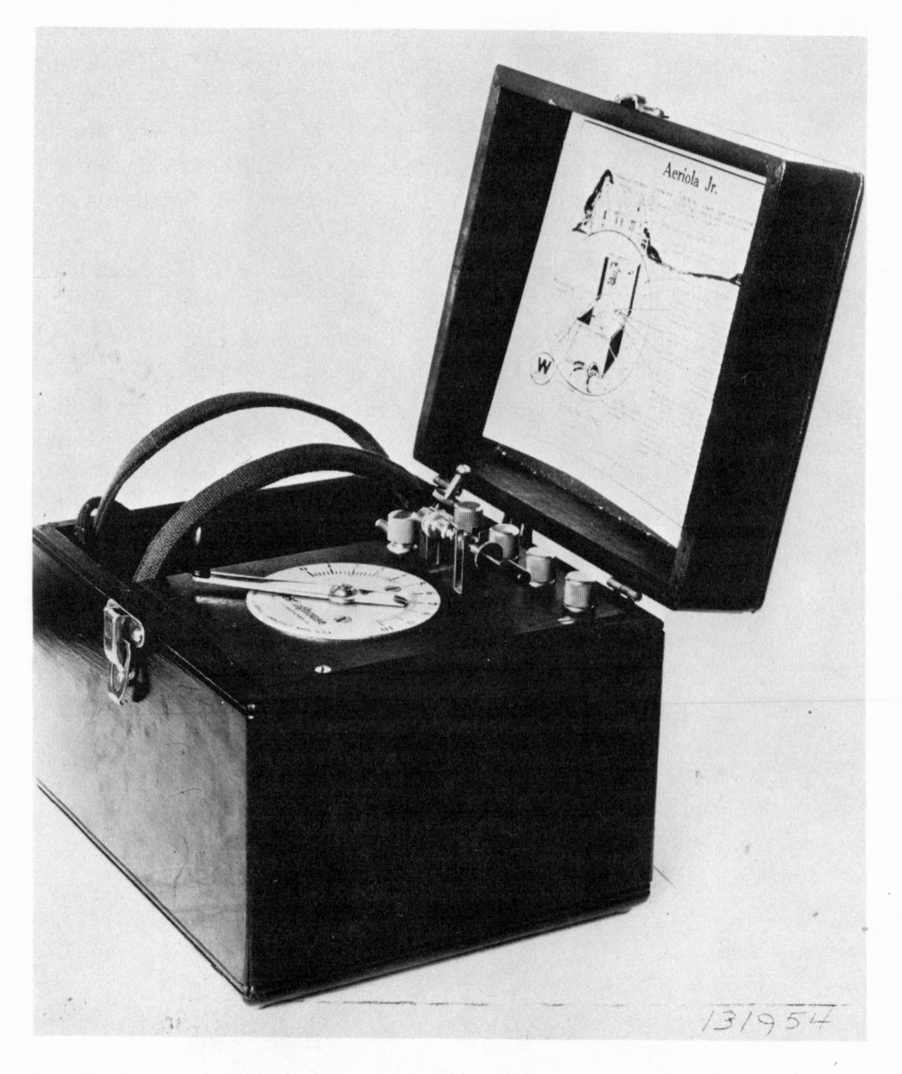

Fig. 32. First popular-priced home radio set, the Aerola Jr. used headphones and cost $25. *Courtesy of Radio KDKA.*

demand for receivers could not be met. One of the greatest booms in American industrial history was under way, launched by the imagination of a radio amateur.

During the next several years manufacturers literally could not keep up with the demand for receivers. General Electric and Westinghouse were joined by RCA in 1922 to become the major manufacturers of sets. Other manufacturers entered the profitable field, and by 1924 several hundred companies were producing receivers. Many of the production facilities were housed in garages or basements. In some cases they were started by radio amateurs, capitalizing on the radio craze sweeping the country. Although RCA owned the patents on Dunwoody's carborundum and Pickard's silicon detector, other minerals, notably galena, could be used. Thus the little manufacturers were not troubled by RCA's patent rights.

Radio engineers for the leading manufacturers realized that if broadcasting were to live up to its potential, better receivers would have to be developed. They looked for ways to improve vacuum-tube circuits both to find new circuitry and to overcome faults in the then almost universal regenerative circuit. The great problem was the tendency of the regenerative receiver to *oscillate,* that is, to become a small transmitting station in its own right, and to send howls and squeals into the neighbors' receivers.

The regenerative receiver had a control that fed back some of the energy in the plate (output) circuit of the tube to the grid, or input circuit. Fed back in exact phase with the input signal, it strengthened the signal to many times that of the original. But if enough energy was fed back, the result was the same as if all resistance in the input circuit had disappeared. Under such conditions, one might expect unlimited current to flow (if there could be such a thing as an unlimited source of current supply). What happened was that the tube "took off" on its own, at the frequency to which its circuits were tuned.

That ability of vacuum tubes to oscillate was what made broadcasting possible—only tube transmitters could put out a pure enough wave for high-quality telephony. But the same oscillation caused a loud squeal in the receiver if the regeneration control was turned up a little too far. It ruined reception in the receiver itself and created interference in all those near it, sometimes being received miles away.

In an attempt to solve the problem, *tuned radio frequency* receivers were introduced. These usually had two stages of amplification tuned to the frequency of the broadcast station. Though there was some feedback from plate to grid circuits, the sets were designed to keep it below the oscillation level. One way was to keep the DC voltage on the grid at a point where the circuit could not oscillate. Another was to put a resistance in the grid circuit. One manufacturer used capacitance in the plate circuit to get the same effect. But all these approaches had their weaknesses. The control that set the voltage on the grid could be used as a regeneration control—and too often was. The sets with grid "suppressor" resistors became very insensitive with age.

The problem was solved by the *neutrodyne* circuit, invented by Professor Louis A. Hazeltine, and introduced on the broadcast receiver market in 1923. The neutrodyne compensated for the regenerative effect due to signal feeding back through the small capacitance between the plate and grid in the tube, by feeding back through a small external capacitance a signal exactly equal and opposite to the one that made the circuit oscillate. Thus more sensitive, more selective receivers that didn't oscillate began to appear on the market in early 1923. The first, brought out by Frank A.D. Andrea, was the famous FADA. It was an instant success, and it and other neutrodynes remained popular for some years, until vacuum tubes with four elements (tetrodes) or five (pentodes) eliminated oscillation in a different way.

Fig. 33. Typical radio receivers of the 1920s and early 1930s. *Courtesy of Westinghouse Broadcasting Co., Inc.*

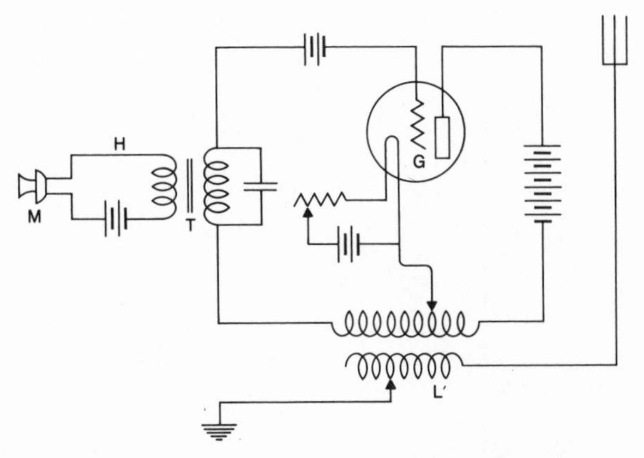

Fig. 34. A simple radio transmitter, such as might be used by an unsophisticated amateur in the early 1920s. *From* Physics, *Hausmann and Slack.*

In 1924 RCA introduced the Armstrong superheterodyne, which soon became the leading type of receiver, and gradually replaced all other types during the next few years.

Figure 34 is the circuit of a simple single-tube transmitter. The sound waves striking the diaphragm of the carbon microphone vary the pressure on the carbon grains in the microphone, reducing its resistance as the compression increases. This generates a varying current through the microphone as the pressure is increased and decreased, following the variations of the sound waves. The varying current in the primary of the transformer T generates a similarly varying current at higher voltage in the secondary. This higher voltage is applied to the grid-cathode input circuit of the triode. The tapped and coupled inductance coils L at the bottom of the figure cause the triode to oscillate and to transfer the signal to the antenna circuit. The frequency can be controlled precisely with a quartz crystal in the grid circuit, and is so controlled in all broadcast stations, citizen's band transmitters, and all other circuits where frequency precision and stability is a must. The voice-frequency voltage on the triode's grid varies the amplitude of the oscillator's output, producing what is called an *amplitude-modulated* radio-frequency signal, which is fed to and radiated by the antenna.

A simple receiver is shown in Figure 35. The radio waves strike the antenna and set up weak currents in it and the antenna coil. The receiver is tuned by varying the capacitance and inductance in the antenna

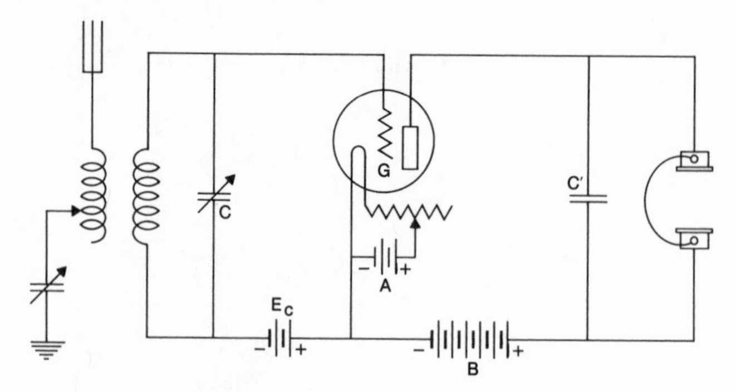

Fig. 35. A simple one-tube nonregenerative receiver.

and the capacitance C in the grid circuit. The weak signal is detected by the vacuum tube and amplified to some extent, and the output is fed to the headphones. More advanced receivers first amplified the radio-frequency impulses with one or two stages of radio-frequency amplification before applying them to the detector.

Armstrong sold his patents for the feedback and superheterodyne circuits to Westinghouse in 1920. These patents became part of the cross-licensing agreements that had been reached earlier when RCA had been formed. Armstrong had retained for himself the right to license these circuits to groups manufacturing amateur sets. At the time, no one could see that broadcasting or amateur radio would mushroom as they did, and the amateur market was not considered important.

Several of the basement businesses catering to amateurs grew rapidly and actually developed better receivers than those made by RCA. As a result, some of these companies were sued, the contention being that they were selling to distributors and could not therefore be considered as selling to amateurs. One of these companies was sued by Westinghouse, but the courts upheld their right to sell to distributors on the ground that at the time Armstrong licensed them, Westinghouse did not contest the licensee's activities.

The success of some of the independents caused the big companies to intensify their research into improved receivers. The superheterodyne receiver finally put RCA in the lead again. Its superiority to other radios of the day was so marked that those independents who wished to continue in business found it necessary to take out licenses from RCA on a royalty basis.

RCA next moved to consolidate its position in the field of vacuum tubes. After the close of World War I the de Forest company had been enjoined from manufacturing and selling Audions until the Fleming patent expired in 1922. De Forest had sold all rights to the Audion in 1916, with the exception of its use in the amateur field. RCA brought suit against de Forest, contending that the de Forest products were being used by others than amateurs. The courts ruled that the spirit of the agreement required that de Forest not use the Audion in transmitting messages either by wire or radio for pay. The key words were "for pay." The landmark decision left de Forest to continue manufacturing vacuum tubes. Unfortunately, he was a better inventor than a businessman, and in 1926 his latest enterprise—the De Forest Radio Company—went into bankruptcy.

The receiver of the de Forest company then sued RCA for violating the Clayton Anti-Trust Act, claiming that a clause in its licensing contract with radio-set manufacturers required licensees to purchase RCA tubes for the sets. Furthermore, it was claimed that the agreements prevented the sale of vacuum tubes for a wide range of uses beyond radio. The suit was won, and several companies filed damage suits against RCA. These were settled out of court. The largest settlement was with the receivership of the de Forest company, which was paid one million dollars.

The setback to RCA was temporary. Research on vacuum tubes and tube circuits was centered in RCA, AT&T, GE, and Westinghouse, with RCA maintaining control of the patents on most new developments. As a result, manufacturers still had to obtain licenses from RCA, with payment on a royalty basis. For a time, RCA also controlled the patents on the best loudspeakers, and companies desiring to manufacture speakers had to obtain licenses from RCA. Thus, the company had a virtual monopoly on the manufacture of radio parts, a situation that would continue for some years.

The mid-1920s saw another bitter and dramatic conflict between de Forest and RCA. RCA officials suspected that de Forest was infringing on their vacuum-tube manufacturing techniques. General Electric had been manufacturing the Langmuir-Coolidge triode, a highly evacuated tube with a thorium-impregnated filament. Heating the filament caused the thorium to rise to its surface, coating it with thorium. This process made it possible to get the same plate current with less power to heat the filament. The tube required an extremely high vacuum, which GE obtained by using the Langmuir mercury-vacuum pump. RCA had a vested interest in this tube because it held the exclusive rights to sell radio equipment in which it was used. Tubes manufactured by the de Forest company had both the extremely high vacuum and the high filament efficiency of the GE tubes, and RCA and GE officials suspected that de Forest was using the Langmuir vacuum pump.

RCA approved the planting of spies—ostensibly ordinary de Forest employees—to gather information to support their suspicions.

De Forest officials—much to the embarrassment of RCA—discovered the spies planted in their midst, and an outcry was raised that was heard throughout the country. De Forest instituted injunction proceedings to prevent RCA from further encroachment, and RCA was immediately branded a villain. The courts enjoined RCA from additional prying. The press was almost unanimous in its condemnation of these industrial warfare tactics.

In spite of the enjoinment, the courts granted RCA the right to use any evidence uncovered in their surreptitious activities. This decision was tantamount to a judgment that the de Forest company had actually acted improperly in its manufacture of vacuum tubes. Yet RCA emerged with the image of oppressor in the struggle, and the reputation of the corporate giant was seriously tainted.

BROADCASTING GROWS

The revolution started by KDKA spread rapidly to other parts of the country, and within two years more than two hundred stations had been licensed. In 1922 the number of licenses issued had grown to 286, and by the end of 1923 a total of 583 stations were authorized to broadcast. The Department of Commerce licensed all applicants because the Radio Act of 1912 has set no standards for issuance of licenses. The drafters of that act could not have foreseen the rapid developments that would take place after World War I. By 1924 interference among stations was so severe than an embargo was placed on further licensing, so that anyone wishing to enter the broadcast field had to buy an existing facility.

The rapid growth of broadcasting resulted in a confrontation between AT&T and RCA. Telephone executives, aware of the great potential of entertainment broadcasting, made an attempt to enter the field. In 1923 AT&T started a campaign to force all broadcast stations to pay it a royalty based on the AT&T contention that the cross-licensing agreements it had entered into with GE gave it the exclusive right to manufacture all radiotelephone equipment, in which category it claimed broadcast transmitters belonged.

AT&T also entered the broadcasting field directly with station WEAF, which it opened in New York in 1922. WEAF carried the first paid commercial program in August 1922, on behalf of a Long Island real estate firm, the Queensborough Corporation. The idea had been conceived by an employee in the commercial department of AT&T.

In January 1923, the first simultaneous hookup of two stations was accomplished, when the programs carried by WEAF were fed by telephone line to Boston and carried by station WNAC. Although the term was not applied until several years later, *network* programming became increasingly frequent in the months that followed.

Although listeners were generally satisfied with these experiments, AT&T engineers were not, and set to work developing cables that could carry high-quality voice transmissions. In the meantime at Westinghouse, Frank Conrad was experimenting with shortwaves, which the amateurs were developing with amazing rapidity; it was thought that they might ultimately be best suited for linking distant broadcast stations.

The first coast-to-coast linkup of stations was in 1924, during the presidential campaign. Twenty-two stations were hooked up to carry a speech by President Coolidge.

AT&T executives refused to make line connections available to any station operating in competition with WEAF. Westinghouse station WJZ was the first affected by the ruling, which prevented it from covering special events from the field; the ordinary method of handling such events was by phoning them from on-the-spot locations direct to the studios. To circumvent the AT&T tactic, WJZ approached Western Union (WU) and made arrangements to use WU telegraph lines where field coverage to a studio was required.

To strengthen its position in the field, the AT&T licensing agreement with broadcast stations stipulated that they could not sell broadcast time to other users. Public opinion weighed heavily against AT&T while negotiations between AT&T and RCA were held. These dragged on while the cross-licensing agreement between AT&T and RCA was tested in the courts. RCA claimed that commercial wireless broadcasting was not included in the agreement with AT&T, which should be limited to radiotelephone transmission for pay only.

The struggle for preeminence in the broadcasting field between AT&T on the one hand, and the manufacturing giants RCA, GE, and Westinghouse on the other culminated in an agreement in 1926 in which station WEAF was sold to RCA by AT&T. As part of the agreement RCA,

GE, and Westinghouse were given the right to manufacture radio receivers and vacuum tubes for public sale, to engage in international wireless telegraphy, and to function in the field of entertainment broadcasting. AT&T was granted the exclusive right to operate in the field of public-service telephone communication.

The 1926 settlement paved the way for the formation of the National Broadcasting Company (NBC). Consisting originally of twenty-four stations in twenty-one cities, the network had its inaugural broadcast in November 1926, when simultaneous programming from the East Coast to as far west as Kansas City was undertaken. The first coast-to-coast broadcast by NBC was carried January 1, 1927, with the airing of the Rose Bowl football game from Pasadena, California.

The Rose Bowl broadcast was carried by a second NBC network, formed when stations around the country pressed the newly formed NBC for affiliation. The networks were designated "red" and "blue" because these were the colors used by NBC personnel in mapping the locations of the transmitters in the networks.

In February 1927, a third network, the Columbia Broadcasting System (CBS), was formed. It consisted initially of sixteen stations. In October 1934, the Mutual Broadcasting System (MBS) was formed with four stations. This basic four-network framework continued until 1943, when an order by the Federal Communications Commission (FCC) required NBC to divest itself of one of its networks. NBC sold the "blue" network, which became known as the American Broadcasting Company (ABC). Regional networks, serving particular areas of the country, have been formed from time to time and have been successful on a smaller scale.

11

Regulating the Broadcasters

The flood of manufacturers and would-be broadcasters to get into the profitable new business started by KDKA was unimpeded either by prudence or law. The only legislation on the books when KDKA started operation was the Radio Act of 1912, which made the secretaries of Commerce and Labor responsible for radio operations. There were, of course, no specific restrictions on broadcasting, and almost anyone applying for a license to broadcast was granted one. Within eighteen months of the KDKA election broadcast, 219 registered stations—and many more unregistered ones—were on the air.

This figure more than doubled by the end of 1924, and by 1927—the year in which broadcasting legislation was first enacted—732 stations were operating. The number of stations broadcasting at some time or another during this period was much higher. In 1924, for example, more than 1,100 stations had been licensed to operate, but more than half of them went off the air for technical or economic reasons. But their impact was felt almost from the beginning.

By 1922, the Department of Commerce had established a wavelength of 360 meters (833 kHz) for the "transmission of important news items, entertainment, lectures, sermons and similar matter." By midwinter of 1922 interference was already a serious problem. Complaints poured in from manufacturers, broadcasters, and the general public—who were spending millions to buy receivers. In the winter of 1922 President Harding instructed Secretary of Commerce Herbert Hoover to convene a conference in Washington, at which representatives of broadcasters, manufacturers, and the general public, including radio amateurs, would be represented. Among the participants at the conference were Hiram Percy Maxim, founder of the American Radio Relay League (ARRL), Edwin Armstrong, and Dr. S.W. Stratton, Chief of the Bureau of Standards.

In his opening address to the conference, Secretary Hoover said:

> We have witnessed in the last four or five months one of the most astounding things that has come under my observation of American life. This Department estimates that today more than 600,000 (one estimate being 1,000,000) persons possess wireless telephone receiving sets, whereas there were less than fifty thousand such sets a year ago. We are indeed today upon the threshold of a new means of widespread communication of intelligence that has the most profound importance from the point of view of public education and public welfare.

The National Radio Conference of 1922 was primarily technical, and treated only engineering problems, including receiver characteristics and operating frequencies. It was decided to make a second channel, 400 meters, or 750 kHz, available for stations operating in areas where the 360-meter wavelength had already been assigned. In addition, minimum powers of 500 watts were required.

Assigning two frequencies instead of one gave beleaguered listeners a little temporary relief. But many still complained because their inexpensive homemade radios (usually a crude crystal detector and a few turns of wire) were not selective enough to tune out one of the two stations if both were in the same area.

As early as 1923 RCA was testing tube transmitters in its transatlantic communication circuits. Two diodes were used to rectify a source of high-voltage AC which was fed to the plate circuits of six triode vacuum tubes, each rated at 20 kW. (Manufacture of high-power vacuum tubes was made possible by the 1922 invention of the metal-to-glass seal at the Bell Telephone Laboratories. Within a year or two of the invention of the Housekeeper—named after its inventor—seal, 50-kW and 100-kW vacuum tubes were made.) These tubes, called *pliotrons* by RCA engineers, generated radio frequency waves that were fed to the antenna circuit of the transmitter.

Radio Broadcast magazine, in its January 1923 issue, described the RCA experiments:

> The installation of these tubes is so simple and the operation so reliable that it is safe to predict that the life of all other high-frequency generators is measured. The tubes used in this first installation have about five thousand

times the power of the small tubes used by amateur broadcasters, but the engineers who are responsible for their development can build tubes of 100 kW or even 1000 kW whenever the demand for them justifies the expense.

The rapid pace of engineering developments in the early 1920s failed to keep up with the unrestricted expansion of broadcasting activities. The chaotic conditions plaguing broadcasters, manufacturers, and listeners alike grew more tumultuous with the passage of time. Two additional National Radio Conferences, in 1923 and 1924, did little to reduce the rate at which confusion increased.

By the time the fourth National Radio Conference was convened in 1925, the broadcast bands had been expanded to include frequencies from 550 to 1,500 kHz, and the maximum power was 5,000 watts. Despite the attempts of Secretary of Commerce Hoover to make some order of the chaos, he was powerless to act. The Radio Act of 1912 had no provisions covering broadcasting, which could hardly have been imagined by its framers. The act had been tested in the courts by broadcasters who fought Hoover's attempts to regulate them. The decision was that the act did not empower Hoover to regulate broadcasting.

The fourth conference broke up in turmoil after failing to agree on anything, including the question of who was to pay the bills of radio broadcasting. This question of financing radio was still a burning issue in the mid-1920s. Herbert Hoover believed that ultimately the radio industry would carry the financial burden of radio programming, but that six or seven national chains would be organized, each supported by segments of the industry.

AT&T, of course, had been selling time over its radio stations, and believed that would be the ultimate means of supporting broadcasting. David Sarnoff, then vice-president and general manager of RCA, advocated endowment of radio stations, similar to that enjoyed by libraries, educational institutions, and museums, because, he reasoned, radio enriched human life and contributed to happiness of the populace and deserved treatment given other such institutions. General Electric Company believed radio would be supported by public contributions or by the licensing of receiving sets, the method being used by the British to help support their broadcasting efforts.

In 1925 there were some 5.5 million radio receivers in the United States. In spite of significant improvements in their sensitivity and selectivity, many listeners were still unable to hear a single interference-free program. After the fourth radio conference, broadcasters who had the money increased power and operating hours as they chose, and Secretary Hoover was unable to do anything about it.

It became evident that some kind of discipline would have to be imposed on American broadcasting by the Congress. In 1926 President Coolidge asked the lawmakers to remedy the confused and tumultuous situation that was wreaking havoc with American radio. The Dill-White Radio Act of 1927 was the result.

The Radio Act of 1927 created a five-member Federal Radio Commission (FRC) to issue broadcast licenses, allocate frequency bands to various radio services, assign specific frequencies to individual stations, and control transmitter power. The same act gave the Secretary of Commerce the power to inspect broadcast stations, to examine and license radio operators, and to assign radio call signs.

The act specified that the FRC could regulate nongovernment radio only. Assignment of frequencies and other operating parameters to government services was placed in the hands of the President, who designated the Interdepartment Radio Advisory Committee (IRAC) to act as advisory body to him. IRAC, formed in 1922, still exists and meets regularly to coordinate U.S. government frequency usage among the various services using the radio spectrum. A member of the Federal Communications Commission (FCC)—which supplanted the FRC in 1934—sits in on IRAC meetings to represent nongovernment broadcasting.

The early efforts of the FRC were devoted almost exclusively to bringing order to the confusion in the broadcast band. Some 150 of the 732 stations operating in 1927 were required to give up their licenses during the first year of the act.

Although the Radio Act of 1927 did much to bring order to broadcasting, there were divisions of power between the FRC and the Department of Commerce, and it became evident in a few years that further legislation was needed. At the request of President Franklin D. Roosevelt in 1933, the Secretary of Commerce appointed a committee to study all nongovernment communication in the United States. This committee recommended that Congress create a single agency to regulate all nongovernment domestic and foreign communication, whether wire or radio. The Communications Act of 1934 established the Federal Communications Commission for the unified regulation and administration of all nongovernment telephone, telegraph, and radio communications operations, domestic and foreign, as well as all communication operations of state and local governments. The FCC began operations on July 11, 1934, as an independent federal agency headed by seven commissioners appointed by the President with the advice and consent of the Senate.

The Radio Act of 1927 had preserved the unique character of American broadcasting by leaving in the hands of commercial nongovernment individuals and companies the responsibility for financing and operating radio broadcasts in this country. Broadcasting in Great Britain had taken a completely different course, and by the time the Federal Radio Commission had met for the first time in March 1927 the charter of the British Broadcasting Corporation had been in existence for some time.

FORMATION OF THE BBC

In early 1919, an experimental wireless telephone transmitter was constructed at Ballybunion, Ireland, for the Marconi Company by Marconi's assistant, H.J. Round. The transmitter, with an input power of 2½ kW, had two vacuum triodes as oscillators and a third as modulator. These vacuum triodes represented the best state of the high-power art at the time. They were made by C.F. Elwell, working for Mullard in England, with quartz envelopes that could stand very high temperatures. Operating on the relatively low frequency of 79 kHz, the transmitter became the first European telephony station to be heard in America.

The Atlantic had been spanned by voice transmission before, but when AT&T engineers had accomplished the feat in 1915, more than 500 vacuum tubes were required. That AT&T transmission, from Arlington, Virginia, was received in Paris.

By early 1920, two higher power transmitters were under test by Marconi engineers at Chelmsford, England. It was customary to test telephone transmitters by reading the names of railway stations, but the Marconi engineers conducting the tests soon became bored and began to recruit company employees with musical talent. Interest in the concerts was widespread, with reports coming from as far as 1,450 miles away, but company officials felt that the future of wireless telephony lay in its commercial application and not as an entertainment medium.

In April 1920, an experimental government station at The Hague, Netherlands, signing PCGG, and operating on 1050 meters (285 kHz), began to transmit concerts that were very well received both in England and on the Continent. British newspapers carried enthusiastic reports of these concerts. As a result Marconi executives reconsidered the possibilities of entertainment broadcasting.

Finally, under the sponsorship of the London *Daily Mail,* a concert given by the renowned Australian prima donna, Dame Nellie Melba, was arranged. Dame Nellie was shown around the Chelmsford plant. As reported by W.J. Baker in his *History of the Marconi Company:*

> Melba was shown the transmitting equipment and then the towering antenna masts, the engineer in charge of the tour explaining that it was from the wires at the top that her voice would be carried far and wide. "Young man," exclaimed Melba, "if you think I am going to climb up there you are greatly mistaken."
>
> Suitably reassured, on that summer evening Dame Nellie stood in front of the microphone (a telephone mouthpiece with a horn made of cigar-box wood fastened to it), and at 7:10 P.M. a preliminary trill came to listeners' ears as the engineers established the best distance between singer and microphone. There followed for the fortunate few who possessed receivers, her magnificent rendering of "Home Sweet Home."

The public responded enthusiastically to Melba's concert, and letters were received from as far away as St. John's, Newfoundland, and Persia. During the months that followed, additional experimental transmissions were carried out, in each instance after Post Office approval had been obtained. In contrast to the uncontrolled activities in the United States, absolute authority for use of communications facilities in Britain rested—as it does to this day—in the hands of the Postmaster-General.

But with entertainment broadcasting increasing rapidly in popularity, the Post Office acted peremptorily and in a manner that was difficult to understand. It withdrew the Marconi license to experiment with wireless telephony on the grounds that these experiments were interfering with "legitimate" services.

Many of Britain's wireless enthusiasts were incensed by this decision, but in spite of the pressures they applied, an application by the Marconi Company to renew transmissions was turned down. During the following year there was renewed pressure by Britain's wireless societies and finally, in 1922, the Postmaster-General granted the Marconi Company authorization to resume wireless telephony tests with a power not to exceed 250 watts. Operating time was restricted to half an hour weekly, and if there was interference to "legitimate" services, total shutdown was to be expected.

On St. Valentine's Day, 1922, experimental entertainment broadcasts were resumed by the Marconi Company on a wavelength of 400 meters (750 kHz), using the call 2MT. The frequency was chosen to avoid interference with other services.

By May 1922 a second experimental transmission from Marconi House on the Strand was authorized, using a maximum power of 100 watts, on a wavelength of 360 meters (833 kHz). Broadcasts from this transmitter, assigned the call 2LO, were an instant success, and were followed by a flood of applications from other manufacturers of wireless equipment. The Postmaster-General held firm, contending that granting licenses to broadcast to all applicants would result in chaos similar to that in the United States. Instead he suggested that a consortium be formed of representatives of interested manufacturers, who would create a single broadcasting entity. The Postmaster-General agreed to consider an application from such a source.

Thereupon, six of Britain's largest manufacturers of radio equipment, the Marconi Company included, formed the British Broadcasting Corporation, Ltd. (BBC). Any manufacturer had the right to purchase shares in the corporation, which was capitalized at £100,000. Operating revenues were

Fig. 36. The first transmitter of station 2LO, London. *Courtesy of the Marconi Company.*

to be obtained from a tax paid by all manufacturers on radio receivers and some accessories sold by them, and by a tax on receivers, to be paid annually by listeners. This receiver tax is common in many European countries to this day.

The British Broadcasting Company was registered in December 1922. Within a year, nine transmitters were operating throughout Britain. The sale of receiver licenses soared. By the spring of 1923 over 87,000 licenses had been issued, with possibly twice that number of listeners owning receiving sets but not paying their tax.

The first receivers approved by the Post Office had crystal detectors and used a coil and sheet of copper, which were moved in relation to each other to change the inductance of the coil and thereby tune the receiver. Later sets, developed by the Marconi Company, had a single vacuum-tube detector with positive feedback from the plate to the grid (regeneration).

The British Broadcasting Company's charter expired at the end of 1926. By then the staff had grown from 4 to 552, and its listeners numbered in the millions. A committee had been set up in 1925 which recommended that the BBC become a Crown corporation, and that it be granted a Royal Charter. Grants thereafter came from Parliament.

Receiver design had developed much more rapidly in the United States. The lag in Britain was probably because the Post Office felt that entertainment broadcasting was not a legitimate service, and therefore actually restricted it. As a result, United States manufacturers reaped the benefits in world markets. It is ironic that this appears to have come about because there was little U.S. regulation until 1927. It is entirely possible that had American officials been able to control U.S. manufacturers as the British Post Office did, American operations might have been similarly restricted.

12

The Amateurs Break Through

While broadcasting was booming in the United States, a revolution in long-distance communication was beginning to develop. New and undreamed-of discoveries that would shrink the globe and bring every corner of the earth into instantaneous communication with each other were being made.

World War I had forced the radio amateurs of the country off the air. As soon as the war was over, they were chomping at the bit, eager to start operating again. The ban on receiving was lifted in April 1919, and finally, in November, after a resolution to lift the ban on transmitting was introduced in Congress, the Navy Department authorized the amateurs to return to the air.

By mid-1920 nearly six thousand amateurs had been licensed by the Department of Commerce, which controlled amateur operations. Consisting largely of returned servicemen who had been radio operators during the war, and of young hobbyists whose imagination had been captured by the new concept of communication without wires, the new group almost immediately filled with their crashing sparks the under-200 meters spectrum they had been allocated.

The amateurs read avidly of new developments, and when the first vacuum tubes became available in the early 1920s, were quick to experiment with tube transmitters. Many were reluctant to give up their full-kilowatt transmitters for equipment that to them seemed to operate at insignificantly low power. But a number had worked with vacuum tubes in the Armed Forces, and knew that a tube operating with a few watts could in many cases outperform spark transmitters of much higher power.

Fig. 37. The first advertisement of the Audion, run in the March 1916 issue of *QST*. *Courtesy of ARRL.*

The continuous-wave (CW) operation of the vacuum tube required entirely different techniques from those used with the interrupted, damped-wave radiations of spark. A spark transmitter operated over a broad band of frequencies; in some cases radiation from a single transmitter covered 50 to 100 kHz or more. Interference was a major problem. The CW transmissions were narrow and sharp, in the order of 1 percent of the bandwidth required by spark. But the receivers built for spark were not generally capable of tuning and holding narrow-band transmissions—drift was a common problem.

During the earlier 1920s, devising new and better receivers became a major occupation of the amateurs. Probably the most famous of the new circuits (Figure 38) was designed by John L. Reinartz, a leading amateur and a pioneer in long-distance amateur communication, using shorter and shorter wavelengths, the equivalent of higher frequencies.

The radio amateur, or "ham," was a very special breed of person, with an insatiable thirst for learning, an irrepressible urge to experiment and to go where none had gone before. Clinton DeSoto, in *Two Hundred Meters and Down,* describes an instance of remarkable amateur ingenuity and perseverance:

> A young lad of seventeen, known to possess an especially efficient spark, CW, and radiotelephone station, was discovered to be the son of a laboring man in extremely reduced circumstances. The son had attended

grammar school until he was able to work, and then he assisted in the support of his family. They were poor indeed. Yet despite this the young chap had a marvelously complete and effective station, installed in a miserably small closet in his mother's kitchen. How had he done it? The answer was that he had constructed every last detail of the station himself. Even such complex and intricate structures as head telephones and vacuum tubes were homemade! Asked how he managed to make these products of specialists, he showed the most ingenious construction of headphones from bits of wood and wire. To build vacuum tubes he had found where a wholesale drug company dumped its broken test tubes, and where the electric light company dumped its burned-out bulbs, and had picked up enough glass to build his own tubes and enough bits of tungsten wire to make his own filaments. To exhaust the tubes he built his own mercury vacuum pump from scrap glass. His greatest difficulty was in securing the mercury for this pump. He finally begged enough of this from another amateur. And the tubes were good ones—better than many commercially manufactured and sold. The greatest financial investment that this lad had made in building his amateur station was 25 cents for a pair of combination cutting pliers. His was the spirit that has made amateur radio.

It was this spirit that in January 1921 had established amateur radio as the fastest cross-country medium of public communication. A relay net-

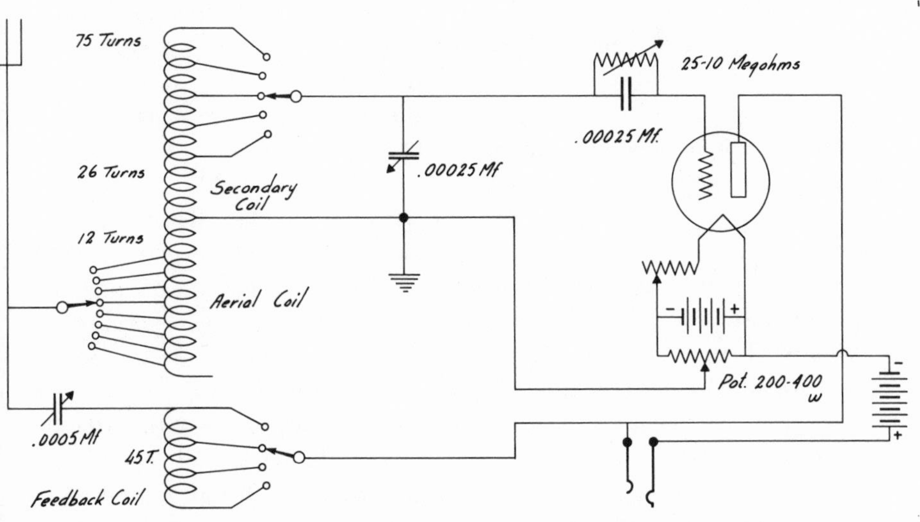

Fig. 38. The Reinartz shortwave receiver circuit. The coil was a "spiderweb"-shaped type. *From* 100 Radio Hook-Ups, *Gernsback Publications, circa 1926.*

work organized by the dominant amateur organization of that day—as well as this—the American Radio Relay League, sent a message across the country and back in six and one-half minutes.

Barely had the excitement of the cross-country feat subsided when RCA announced that it was making commercially available two transmitting tubes, the UV-202 and UV-203, the former a 5-watt oscillator selling at $8, the latter a 50-watt tube for $30. General Electric in the same year offered a still larger triode selling for $110 and operating at the relatively high power of 250 watts.

The death knell of spark had been sounded.

The availability to amateurs of vacuum tubes resulted in a renewed surge of interest in the hobby, and by mid-1921 the Department of Commerce had licensed some eleven thousand amateurs. The increase in the number of active operators, added to the growth of other services, created more problems in radio.

Funds to enforce the Radio Act of 1912 were not available; the provision of that law to restrict amateur operations to 200 meters (1.5MHz) or below was not always applied. The growth in importance of radio after World War I, however, made it necessary to watch the spectrum space more closely than before. Mindful of the potential dangers to amateur operation if the regulations were not adhered to, the ARRL organized a self-policing operation designed to keep amateurs operating on the legal wavelength of 200 meters. The free-for-all period, in which the hams had operated pretty much where they pleased, was voluntarily brought to an end. Had it not been for the efforts of the ARRL, it is likely that pressure to put an end to amateur radio by legislative means would have been great enough to practically end the hobby.

The idea of transmitting amateur signals across the Atlantic had been proposed by Hiram Percy Maxim, founder of the ARRL, as early as 1914, and the concept had been discussed on several occasions after that. But it was not until February 1921 that the first organized attempts were made. A handful of American amateurs had arranged to transmit prearranged signals, for which several hundred British amateurs would listen. Prizes were offered by manufacturers of radio equipment on both sides of the Atlantic. Most of the British amateurs participating in the tests used regenerative receivers, which tended to oscillate and become small transmitters. As a result, unwanted radiation was produced, creating considerable interference and preventing reception of the stateside signals. Other factors, including interference from commercial services, high noise levels, and uncertainty of receiver calibration, resulted in failure. It was decided to repeat the tests later in the year and to send an American amateur to England with the latest available receiving equipment.

Accordingly, Paul Godley, 2XE, a foremost receiving expert, was sent to England with a superheterodyne and regenerative receiver. By early December 1921 Godley was set up on the bleak and windy west coast of Scotland at Ardrossan. Within a matter of days, Godley, as well as several British amateurs, had picked up more than thirty American stations, including one as far west as Pittsburgh. Godley received these signals on frequencies ranging from 1.1 to 1.5 MHz (200 to 270 meters).

The lead story in the January 1922 edition of the ARRL's *QST* Magazine began: "Oh, Mr. Printer, how many exclamation marks do you have? Trot 'em all out; we're going to need them badly, because WE GOT ACROSS!!!!!!"

Although nine spark transmitters had been logged by Godley during his ten-day stay at Ardrossan, the American amateurs who used tubes got across with significantly less power than did the sparks, and all but the most hard-headed began to concede that CW was superior to spark.

The dramatic transatlantic tests failed to overshadow a crisis for amateur radio that developed during the last few months of 1921 and culminated in the first National Radio Conference of 1922. The broadcasting boom had captured the imagination not only of the lay public, but some amateurs as well. Visualizing themselves as budding radio personalities, many amateurs began to broadcast their own programs, often on considerably longer wavelengths than the 200 meters to which they were entitled. Literally hundreds of amateurs were putting on programs, many entirely without commercial intent. By January 1922 interference to broadcasting was so severe that the Secretary of Commerce forbade broadcasting by amateurs, and required that all broadcast stations obtain commercial authorizations before being allowed to transmit.

Although there was some question of the validity of the amendments to the Radio Act of 1912 at the time, the amateur community as a whole wisely chose to abide by the new regulations. Some interference continued, however, and at the First National Radio Conference in 1922, a silent period was imposed on all amateur stations from 8:00 to 10:30 P.M. daily, and during Sunday-morning church services. Amateurs were assigned the bands between 150 and 275 meters, with the region above 200 meters to be shared with training schools. The conference also defined a radio amateur for the first time as ". . . one who operates a radio station, transmitting or receiving, or both, without pay or commercial gain, merely for personal interest or in connection with an organization of like interest."

Under the leadership of the ARRL, amateurs throughout the country complied with the regulations as amended at the conference, which assured that amateur radio was not only preserved, but strengthened, because their activities had been defined for the first time. The stage was now set for the greatest of all amateur achievements—establishment of low-power two-way long-distance communication between remote areas of the world and the discovery of the usefulness of the shortwaves.

In early 1922 over fourteen thousand amateur stations were licensed, and with the National Radio Conference having supported their continued existence, amateurs set about to expand their horizons. A third series of transatlantic tests, organized in 1922, saw 316 American stations received in Europe. Signals from every part of the United States, including the West Coast, were heard. Equally significant, during these tests French and British amateurs were heard in the United States for the first time. The implication was clear: amateur two-way communication across the Atlantic was possible!

While transatlantic testing was capturing headlines in American publications, merchant seamen operating on American ships had been logging American stations as far west as the coastal waters of Japan and China, to distances as great as 6,000 miles. In the summer of 1923, amateurs were opening up the radio paths between the United States and Australia and New Zealand. During the autumn, some one hundred U.S. stations were logged "down under." The distances ranged to 10,000 miles, and were climbing. The first attempts to establish two-way communication between the United States and France were made in January 1923, and failed. The French station could be heard in the United States, but no American could be heard in France.

A further modification in the bands allocated to amateur radio was made at the second National Radio Conference in March 1923. The amateur bands were confined to the region 150–200 meters (1.5–2.0 MHz), the wavelengths between 200 and 275 being withdrawn. The upper portion of the amateur bands was limited to the fast-dwindling spark transmitters.

FIRST TWO-WAY AMATEUR TRANSATLANTICS

One of the high points in the history of amateur radio—in all radio, for that matter—was recorded in the fall of 1923. In January of that year a series of transatlantic tests had failed, even though a French station, 8AB, had been heard regularly. After the tests the operator of 8AB, Leon Deloy, visited the United States to learn all he could about American amateur operations. He purchased American equipment, consulted with communication experts, and studied American techniques with the avowed intention of being the first European to effect two-way communication with the States.

Deloy returned to France in the autumn of 1923, set up his newly purchased equipment, and began testing in October. By mid-November he was ready. He cabled the ARRL that he would commence transmitting on 100 meters (3.0 MHz) at 9 P.M. November 25, 1923. The CW signals of 8AB were audible the first night of the tests, and although scores of amateurs attempted to communicate with him, none succeeded. Nor was it any better the next night.

However, on the night of November 27, after transmitting for one hour, and asking for an acknowledgment, Deloy sent back the electrifying news—he had received answers to his signals from John Reinartz, W1XAM, and Fred Schnell, W1MO. He replied to the American stations, then signed off with the meaningful phrase: "This is a fine day!"

For the first time in history, radio amateurs had communicated with each other from opposite sides of the Atlantic! Even more remarkable, communication was on the shortest waves—100 meters, or 3 MHz—yet to span the Atlantic.

The most hectic twelve months in amateur history followed. Although few amateurs realized it at the time, a new era in long-distance communication had dawned, as one by one records for distance were shattered almost daily. The first two-way Anglo-American amateur contact was made in December, and in the same month voice signals were transmitted in both directions, the American transmitter accomplishing the feat with two 5-watt tubes. This in itself was a step forward, since radiotelephones have a shorter range for a given power output than the CW transmitters used in earlier tests.

Most of the two-way transatlantic communication had been in the "useless" bands between 100 and 118 meters. Until then, little thought had been given to how these waves propagated, or whether the use of shorter wavelengths was a factor in enhancing the transmission. The success was at first attributed to lower interference in the "clean" range far below the crowded 200-meter band and to more efficient operation of antennas that were large for the frequencies they were using. The success in the 100-

meter range prompted the ARRL, through its magazine *QST*, to call for a general movement down in wavelength, to clear some of the severe congestion in the 200-meter amateur band.

So early tests in the region below 100 meters were conducted not so much because amateurs felt they had discovered a portion of the spectrum with significantly different propagation characteristics, but because this region was relatively free of interference. Congestion in the higher bands was the primary reason for the move to higher frequencies (which was still always called a move downward) and for the experimental efforts that culminated in an altogether unexpected series of discoveries. That uncharted region below 200 meters was to offer opportunities no amateur of the early 1920s dreamed of!

There should be nothing surprising about that. For twenty years some of the most eminent scientists, Marconi included, felt that propagation improved as wavelength increased. This was, to some extent, true. With plenty of power behind them, the long waves were reliable. As the frequencies became higher, they became less reliable, and—particularly during business hours—tended to fade out altogether. Not only the region from 200 meters down, but the whole of the present broadcast band and a considerable band below it in frequency were considered unsuitable for reliable commercial radiotelegraphy.

Meanwhile, great strides were being made by those amateurs who confined their activities to operation on this continent. They made up a large majority of the 16,000 amateurs with licenses to operate, and as they moved up to higher frequencies, they developed increasingly sophisticated transmitting and receiving equipment, using vacuum tubes, more efficient tuners, improved antenna systems, and newly designed amplifiers.

By 1924 the amateur program of linking the world by radio had expanded in two new directions, as two-way contacts were established between North and South America and between South America and New Zealand. In their quest for new horizons the hams developed a shorthand language all their own. Because those long-distance contacts were sometimes ephemeral, abbreviations had to be used wherever possible to economize on time. Thus distance became *DX*, operator became *OP*, worked (communicated with) was abbreviated *WKD*, *YL* stood for young lady, and *XYL* for wife (ex-young lady). These were combined with "phrases" from the older wire telegraph: *73* meant best regards, *88* love and kisses (the "good numbers" of the present-day CB enthusiasts).

By mid-1924, interest in the shortwaves which the amateurs had pioneered and developed had spread throughout the world. Amateurs were asked to cooperate with the governments of Canada, Italy, France, and the United States in a series of tests designed to further their research into the shorter wavelengths.

While the expansion into the uncharted portion of the short-wave region was going on, the ARRL was negotiating with the Department of Commerce for authorizations at still higher frequencies (shorter wavelengths) in the spectrum. In July, amateurs were authorized to use one or more of the following bands if prior approval in the form of modified licenses was obtained: 75–80 meters (3.75–4 MHz), 40–43 meters (70–75 MHz), 20–23 meters (13–15 MHz), and 4–5 meters (60–75 MHz). Spark transmission in these new bands was banned. Only CW could be used. With these authorizations granted, amateur radio stood, in August 1924, on the threshold of its greatest achievement, the development of shortwave radio.

The accomplishment of the amateurs in using the wavelengths below 200 meters was all the more remarkable because at the time no existing technology could theoretically make long-distance communication in these bands work, and no known laws of science could account for the propagation of radio waves in the regions used by the amateurs over such long distances.

Little wonder, therefore, that many eminent scientists doubted that the wavelengths below 200 meters would be of much use for long-distance communication. Clinton DeSoto, in *Two Hundred Meters and Down*, quotes an eminent radio engineer who was interviewed in 1921 by the editor of QST Magazine. The interview, held before the transatlantic tests, is typical of much of the thinking of the day:

> "It can't be done," he announced dogmatically. "Why," he explained, vest-pocket slide-rule in hand, "the number of amperes that with a kilowatt input can be erected at the base of a 200-meter transmitting aerial of optimum effective height simply isn't capable of inducing the minimum required microvolts-per-centimeter of receiving aerial length to produce a signal of unit audibility at anything like that distance!"

The belief that it couldn't be done was based on two common misconceptions. First, the reflection characteristics of the ionosphere, which had been postulated some twenty years earlier, had not yet been proved. There were still doubters who were skeptical not only of its characteristics but of its very existence. Second, no allowance was made for the development of the vacuum tube, as amplifier, as oscillator, and as detector. There had been plenty of evidence of the potential of this incredible little glass bulb, but most people were not able to see the enormous possibilities the miraculous little device offered.

The amateur, confined to the region of 200 meters and down, became a specialist, and through his journal, *QST*, wrote and informed his fellow amateurs of the wonderful, undiscovered world that lay in the as yet unsullied region of the electromagnetic spectrum. As early as 1922, in an article entitled "Radio Below 200 Meters," amateur Boyd Phelps wrote in *QST* of an experiment in which a 100-watt tube transmitter was tuned and operated as low as 35 meters. Measurements indicated that at this low wavelength the antenna was radiating energy. Of course, no one was on the receiving end, but Phelps urged his fellow amateurs to move downward.

And move downward many of them did. Phelps continued running tests with other amateurs on wavelengths on the order of 100 to 135 meters. Among the pioneers joining in these tests was 8XK, Frank Conrad, whose earlier experiments had led to the first KDKA broadcasts. Within one year of the two-way transatlantics, radio signals were transmitted and received between the antipodes, when a New Zealand amateur and one from London carried on two-way communication for more than an hour. The amateur had gone as far as he could on this earth, and a prize was offered for communication with Mars (probably long forgotten before the first such communication took place).

As amateurs gained experience in 80-, 40-, 20-, and 5-meter band operation, they learned of important—even startling—differences in their propagation characteristics. As early as 1925 their experiments in the short-wave spectra contributed heavily to our knowledge of VHF propagation.

One phenomenon, *skip distance*, was immediately noted. A signal from, for example, an 80-meter transmitter would be heard up to 20 miles or so, then weaken and disappear. But, about 800 miles from the transmitter, it would appear again at good strength. Nothing could be heard in the region between 50 and 800 miles. It was evident that the signals were making a "jump" somewhere and somehow.

The behavior of signals in the 20-meter band astounded the amateurs. It had become a matter of gospel that signals propagated much better in the dark. But 20 meters reversed the situation, and hams found themselves communicating with each other across the continent at noon, with insignificantly low power. At night 20 meters was useful only for short distances—up to 20 miles or so.

The 5-meter band, the amateurs learned, had another set of characteristics. It seemed to be useful for distances of only a few miles at any time. The hams began to get an inkling of line-of-sight transmission, now accepted so matter-of-factly by FM radio listeners and television viewers.

Although amateurs were first to open the shortwave region of the spectrum to radio communication, they did not stop there. An amateur who was also a scientist first described some of the characteristics of that reflecting layer above the surface of the Earth, first postulated at the turn of the century by two scientists, Arthur Kennelly of Harvard, and Oliver Heaviside, in England. Writing in the May 1924 issue of *QST* Magazine, Dr. A. Hoyt Taylor of the Navy Department, a prominent amateur and shortwave pioneer, wrote: "I must conclude that there is a complete reflection of these waves at some upper and probably ionized layer of atmosphere."

Another amateur, John Reinartz, suggested in the April 1925 issue of *QST* that the reflection or refraction in the Kennelly-Heaviside layer was a function of the frequency. Much of the groundwork for the explanation of the characteristics of radio waves and the way they propagate was laid by these two amateurs, and by K.M. Jansky, Jr., who suggested that the distance to which radio waves traveled along the ground was also a function of frequency—the higher the frequency, the greater the ground attenuation. Signals received beyond the point of ground-wave range were those propagated by the Kennelly-Heaviside layer.

By 1925, the amateurs had therefore contributed their most important finds to global communication; they had opened the shortwaves on a worldwide basis and had explained, in a rudimentary way, the propagation characteristics of some of their signals. It remained for the scientists to step in and explain fully how the magical phenomenon of long-distance communication without wires worked, and why its characteristics were so different at different frequencies.

13

Radio Wave Propagation

In 1911, some ten years after his historic transatlantic experiment, Guglielmo Marconi said in an address before the Royal Institute of Great Britain:

> Although we have—or believe we have—all the data necessary for the satisfactory production and reception of electric waves, we are yet far from possessing any very exact knowledge concerning the conditions governing the transmission of these waves through space—especially over what may be termed long distances. Although it is now easy to design, construct, and operate stations capable of satisfactory commercial working over distances up to 2,500 miles, no clear explanation has yet been given of many absolutely authenticated facts concerning these waves.

Although James Clerk Maxwell had brilliantly formulated the mathematics of electromagnetic wave propagation through space some fifty years earlier, the mechanics by which radio waves traveled over the Earth to remote points remained a mystery. Light, another form of electromagnetic radiation, traveled nearly in straight lines, and it had been shown that wave diffraction (the phenomenon that results in sound waves traveling over the top of a hill or around the corner of a building) could not account for the bending of radio waves around the surface of the Earth.

Attenuation along the Earth's surface made it virtually impossible for the energy to travel along the ground for anything nearly approximating the long distances that had been covered. The strength of the signals received at distances of thousands of miles were far too great to account for by travel along the ground.

More and more attention, therefore, was directed toward the theories of two scientists, who independently at the turn of the century had postulated a shell that could conduct radio energy surrounding the Earth. Oliver Heaviside, in England, and A.E. Kennelly, in the United States, independently suggested, in discussing Marconi's transoceanic transmission, that there must be a reflector overhead that could bring the waves down to Earth at a distance, and that this reflecting layer must be formed by free electrical charges in the upper atmosphere. Heaviside made his comments in the *Encyclopaedia Britannica,* and Kennelly in the *Electrical World,* both in 1902. The idea was further detailed by Professor G.W. Pierce of Harvard in 1910 and W.H. Eccles in 1912.

The first rough measurements of this "radio reflecting layer" were made by Lee de Forest and Leonard F. Fuller between 1912 and 1914. Working with the continuous-wave (arc) transmitters of the Federal Telegraph Company in San Francisco, de Forest found that signals might fade at one frequency while increasing at a nearby one (the arc transmitter was keyed by shifting its frequency slightly from the "working wave," so that it could oscillate continuously). This suggested that the direct wave propagated along the surface of the sea might be combining alternately in and out of phase with a "sky wave," a possibility that Professor Pierce had mentioned. Publishing his observations in the first issue of the *Proceedings* of the Institute of Radio Engineers, de Forest said, "If the reflecting layer is halfway between the stations its height is 62 miles. . . ." His colleague, Leonard Fuller, continued the studies during 1913 and 1914 and published a more detailed report in the *Transactions* of the American Institute of Electrical Engineers.

In the 1923–24 period, E.O. Hulburt and A.H. Taylor of the U.S. Naval Research Laboratory attempted a new approach. Analyzing a large assemblage of data on successful radio communication, they undertook to describe the worldwide properties of the reflecting layer. While they were studying the layer in this manner, two British scientists, Edward Appleton and M.A. Barnett, conducted a series of experiments in which the frequency of a transmitted wave was varied, and the strength of the received signal monitored a short distance from the transmitter. Appleton found that the strength of the received signal varied with frequency. This indicated that at certain frequencies interference patterns were being set up between the part of the signal traveling along the ground and the part of the signal that traveled upward and then was returned to Earth. The difference in time between the signal being received direct and the one reflected from the upper atmosphere could be determined by the number of maxima and minima in signal strength for various frequencies and by the distance between the transmitter and the receiver. Appleton deduced from these experiments that the reflecting layer was about 100 kilometers above the Earth. This agreed closely with de Forest's earlier figure of 62 miles.

At about the same time two American physicists, Gregory Breit and Merle A. Tuve, used a more direct method of demonstrating the existence and height of a reflecting layer. Breit and Tuve transmitted pulses of energy straight up, and, with receiving equipment set up nearby, measured the time it took for the echoes to be received. Three and sometimes more bursts of energy were received on their measuring equipment, the first from the wave traveling directly from transmitter to receiver, the second from a conducting region about 60 miles above the Earth, and third from a second, higher conducting region in the upper atmosphere.

Later, Breit and Tuve varied the frequency of the transmitted pulse and discovered that there was always some upper frequency, which they called the *critical frequency*, at which reflections would cease and no signal would be returned.

The pioneering experiments of Appleton, Breit, and Tuve provided the first conclusive evidence of the existence of a conducting region that girdled the Earth, and gave a clue to the nature of its physical characteristics.

The Breit-Tuve experiments were repeated at numerous locations throughout the world. It was soon learned that the critical frequency varied according to the time of day, season, and geographical location. These experiments also indicated that the conducting regions in the upper atmosphere were influenced by radiation from the sun.

Additional evidence linking the characteristics of the conducting layers with solar radiation was obtained during the solar eclipse of 1927. A sharp decrease in critical frequency was observed during the period of totality. The scientists concluded that the reflecting layers were primarily influenced by ultraviolet radiation from the sun. Later experiments confirmed that ultraviolet radiation and—to a lesser extent—X-rays are the primary agents in the formation of a conducting region above the Earth.

The conducting region was first referred to as the Heaviside layer, the Kennelly-Heaviside layer, or the Appleton layer. But Sir Robert Watson-Watt, an associate of Appleton's and one of the original developers of radar, called the entire region collectively the *ionosphere*, and this term has been adopted internationally.

FORMATION OF THE IONOSPHERE

The Earth's atmosphere is composed mainly of oxygen, nitrogen, hydrogen, and helium. At distances relatively close to the Earth, these gases occur in a rather homogeneous mixture, and are kept uniformly mixed by the action of the weather. In the upper regions of the atmosphere, weather effects disappear. The gases tend to become separated, with the lighter ones rising higher than the heavier ones. There is therefore a tendency for particular gases to exist at certain levels. Gas density increases as the lower levels are approached. Ultraviolet intensity decreases as density increases, because of cumulative absorption.

Like all matter, gases are composed of atoms. The *atom* is a fundamental unit of matter, made up of a positively charged nucleus surrounded by units of negative electricity, or electrons. When each atom has its normal number of electrons, it is said to be *neutral*.

The ionosphere is formed when photons of ultraviolet light and X-rays from the sun bombard these various gas atoms high in the upper atmosphere. The atoms absorb energy from these rays and are set into such a state of agitation that electrons are dislodged from the formerly neutral atoms. This produces free electrons and atoms that have lost one or more

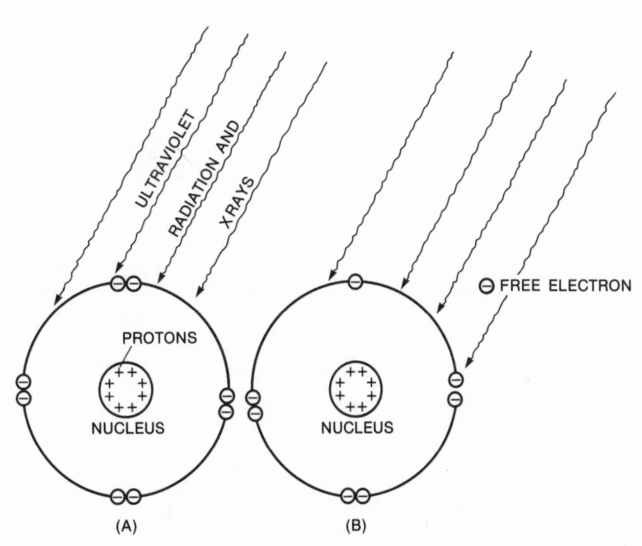

Fig. 39. The effects of outside exciting forces on an atom are shown in this highly simplified drawing.

electrons, or *ions*. Because it has lost negative charge, the net charge on the previously neutral atom is now positive; it is therefore referred to as a *positive ion* (see Figure 39).

Because gases of different densities are encountered at different heights above the Earth by the descending ultraviolet rays, several distinct layers of ionized gas are formed in the upper atmosphere. The free electrons within the layers can move independently of the surrounding ions, so each layer of ionized gas acts as a metallic conductor and can therefore reflect radio waves to Earth.

If the strength of the ultraviolet and X-radiation decreases, the free electrons begin to recombine with the ions in the various ionospheric layers. This process of *recombination* occurs primarily during the hours of darkness, when the ionosphere has been cut off from solar radiation.

THE IONOSPHERE'S STRUCTURE

As solar radiation penetrates the Earth's atmosphere, it begins to ionize the rarefied layers of gases it encounters. As it penetrates deeper, gas density, and therefore ionization density, increases. At some level, the radiation decreases because of absorption as it penetrates deeper into the atmosphere. Beyond this maximum point, ionization begins to fall off. Thus a region is formed in which ionization is a maximum at a certain level, falling off both above and below it.

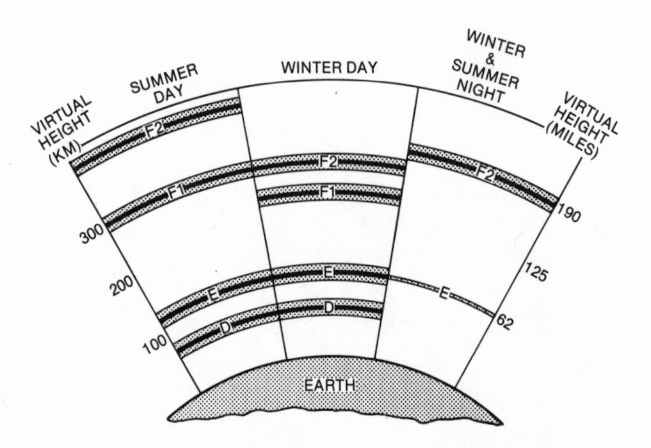

Fig. 40. The height of the ionospheric layers varies with the time of day and the season.

The range of ultraviolet radiation from the sun covers a relatively wide band of frequencies. Since the gases comprising the upper atmosphere respond to different frequencies in the ultraviolet spectrum, ionization occurs at several different levels, or layers, between approximately 30 to beyond 400 miles above the surface of the Earth.

These ionized regions are usually referred to as "layers." However, they are not separated from one another as the term would indicate. Each region, or layer, overlaps, forming a continuous but nonuniform area with at least four levels of peak intensity, designated D, E, F_1, and F_2.

It was Appleton who assigned letters to the layers of the ionosphere. He used the letter *E* to label his first discovery. When he found a second, higher reflecting layer, he used the letter *F*, and when a third, lower layer was discovered, he labeled it *D*. Appleton said that this left other researchers an adequate number of letters to allocate to any undiscovered layers.

THE LAYERS

The D layer appears during daylight hours only, at an average altitude of 40 miles (64 km). The ionization in this region is low compared to the higher layers, and is maximum when the sun is overhead, disappearing almost entirely during the hours of darkness. During solar storms, when solar radiation increases tremendously, the D layer absorbs radio waves in the shortwave region of the spectrum, preventing the reflection of radio waves and thereby causing signal failure on long-range radio circuits.

Like the D layer, the E layer exists primarily during the daylight hours, disappearing almost completely at night. Its height remains relatively constant throughout the year, with maximum density at an altitude of about 60 miles (96 km).

The F layer is the most important one. It is the region from which most long-distance shortwave radio signals reflect. Its height varies from approximately 100 to 300 miles above the surface of the Earth, depending on time of day and season of the year.

During the daylight hours the F layer tends to break into two regions, F_1 and F_2. The F_1 layer behaves like the D and E layers, with maximum density around local noon. Because recombination in the F_2 layer is relatively slow, that layer continues to exist around the clock, making long-distance nighttime communication by reflection possible (Figure 40).

WHAT ARE RADIO WAVES?

Electromagnetic waves consist of rapidly varying electric and magnetic fields which travel through space at right angles to each other. Radio waves, like other forms of electromagnetic radiation (light, X-rays, gamma rays, infrared radiation, and ultraviolet light), travel at about 186,000 miles (300 million meters) per second in free space.

If we examine an electromagnetic wave, we find that it is made up of two distinct components—the electric field and the magnetic field.

Figure 41 illustrates the electric and magnetic component of a wave traveling outward from point O toward the right. The magnitude of both the electric and magnetic fields varies continuously. This figure shows only the E (electric) component of the wave at different times. We can see that the electric field has changed from a maximum in one direction, through zero at point O′, to a maximum in the other direction, through zero again, and back to the original magnitude and direction shown. This change in E from a given condition through all its intermediate values to its original condition is a *cycle*. During the cycle, every value and direction of E has been traversed. If we draw a smooth curve joining every possible value of E, we get a *sine curve*, as shown, and E is said to vary sinusoidally. The magnetic (H) component of the radio wave, which can be represented by an arrow coming out of the paper toward the reader, also varies sinusoidally.

Figure 41 illustrates the simplest type of radio wave—one in which the E and H components are constant and in which the relative magnitude and direction remain fixed. Actually, very complex relationships between E and H are possible.

In Figure 41, the E component varies in a vertical direction. If this variation is with respect to the Earth, the wave is said to be *vertically polarized*. If the component varies horizontally, the wave is *horizontally polarized*.

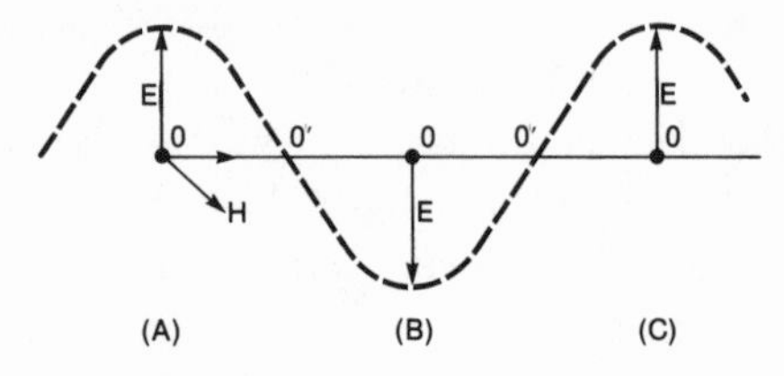

Fig. 41. Electronic and magnetic components of a radio wave. The electric component is marked E; the magnetic component can be thought of as pointing toward the reader.

The number of cycles of a radio wave passing a fixed point in a given time is the *frequency*. Thus, if the time taken for a complete cycle is 1 second, the frequency is 1 cycle per second.

By international agreement, a frequency of 1 cycle per second is defined as a Hertz. Because radio waves oscillate at extremely high frequencies, it is more convenient to use kiloHertz, megaHertz, and gigaHertz:

1000 cycles per second = 1 kiloHertz (kHz)
1000 kHz = 1 megaHertz (MHz)
1000 MHz = 1 gigaHertz (GHz)
1000 GHz = 1 teraHertz (THz)

The relationship between the velocity of an electromagnetic wave, c, its frequency, f, and its wavelength, λ (lambda) is $c = f\lambda$, where $c = 300{,}000$ kilometers per second, f is in kHz, and λ is in meters.

The relationship between the velocity of a radio wave, its frequency, and its wavelength is:

$$\text{Frequency (in KHz)} = \frac{300{,}000 \text{ km/sec}}{\text{wavelength (in meters)}}$$ Thus,

$$f = \frac{300{,}000}{\lambda} \qquad (1)$$

or, transposing, $$\lambda(m) = \frac{300{,}000}{f \text{ (in KHz)}} \qquad (2).$$

A picture of the relationship between frequency, wavelength, and velocity of a radio wave may help understand this concept. Figure 42 depicts

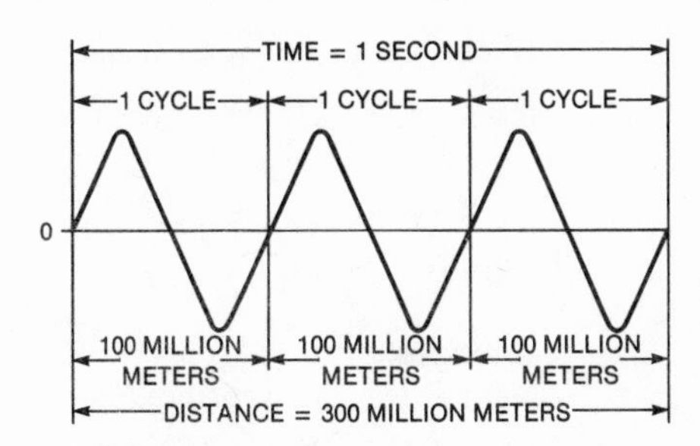

Fig. 42. Three cycles of a 3-megaHertz radio wave.

a radio wave that has left its antenna 1 microsecond ago and has completed 3 cycles. It has traveled 300 meters.

Since 3 full cycles have been completed, the length of the wave, or the wavelength (one wave is radiated for each complete cycle), equals 300 million meters divided by 3 million (the number of cycles that would occur in a second). This equals 100 meters.

To convert frequency to wavelength, or vice versa, one of the quantities f or λ must be given. For example, suppose we have a wavelength of 10.15 meters, and we want to know its frequency. Using equation (1), we get

$$f = \frac{300{,}000}{10.15} = 29{,}557 \text{ kHz}$$
$$= 29.557 \text{ MHz}$$

Given a frequency of 7063 kHz and asked to find the wavelength, we would use equation (2):

$$\frac{300{,}000}{7063} = 42.47 \text{ meters}$$

CHARACTERISTICS OF RADIO WAVES

The radio communications spectrum is subdivided by international agreement into several frequency bands:

BAND NUMBER	FREQUENCY RANGE (LOWER LIMIT EXCLUSIVE, UPPER LIMIT INCLUSIVE)			DESIGNATION
2	30 to	300 Hz	Extremely low frequencies	ELF
3	300 to	3,000 Hz	Voice frequencies	VF
4	3 to	30 kHz	Very low frequencies	VLF
5	30 to	300 kHz	Low frequencies	LF
6	300 to	3,000 kHz	Medium frequencies	MF
7	3 to	30 MHz	High frequencies	HF
8	30 to	300 MHz	Very high frequencies	VHF
9	300 to	3,000 MHz	Ultra high frequencies	UHF
10	3 to	30 GHz	Super high frequencies	SHF
11	30 to	300 GHz	Extremely high frequencies	EHF
12	300 to	3,000 GHz or 3 THz	No present designation	

The propagation characteristics of radio waves differ depending on frequency. All radio waves have certain things in common. For example, when a radio wave leaves a source it normally travels outward in all directions. For communications purposes, such radio waves can be classified in two categories, ground and sky waves.

The ground wave is the portion of the transmitted energy that is influenced by the Earth and its surface features. There are two components of the ground wave: the portion of the energy that travels along the Earth's surface and is guided by it is the *surface wave.* The portion of the energy that travels in a straight line from the transmission point to the receiving site is the *direct wave* (Figure 43). The direct-wave path is sometimes called line-of-sight transmission; it is generally limited to relatively short distances in which the transmitter and receiver are in sight or almost in sight of each other. The distance the surface wave travels increases at longer wave lengths (lower frequencies). In bands 4 and 5, for example, hundreds, perhaps thousands of miles can be covered with the surface wave, depending on the radiated power. In band 6 these distances are reduced to hundreds of miles, and by the time band 7 is reached the direct wave travels farther than the surface wave, because of attenuation along the Earth's surface.

Band 6 includes the medium-wave portion of the spectrum—the standard broadcast band in this country. During the day, coverage in this band (550–1550 kHz) is limited to the ground wave and is generally of the order of 50 to 75 miles, depending on the transmitter power and the antenna characteristics. At night, the coverage range increases significantly, and interference levels in most rural areas rise during the evening hours. This is because at night medium waves are reflected from either the residual E layer, or from the F layer, depending on the degree of ionization in these layers. This is why stations in the American broadcast bands that are never

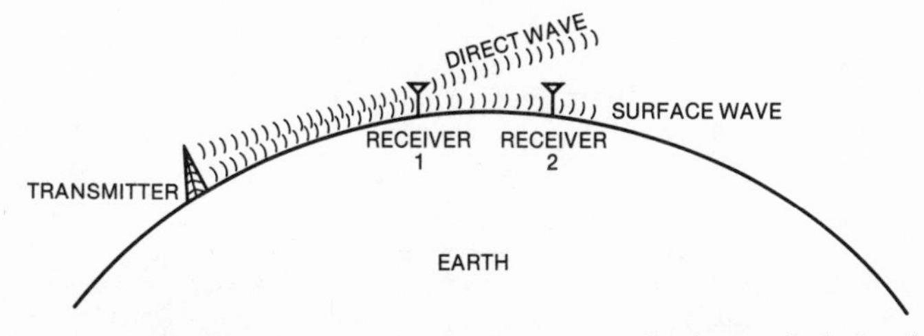

Fig. 43. Surface and direct waves. The direct wave is also bent a little by the Earth, even at higher frequencies, and so can be received at distances greater than line-of-sight.

heard by day are easily receivable in many locations during the evening hours.

The most important component of all but the longest radio waves is by far the sky wave—that portion of the transmitted energy that travels upward, away from the surface of the Earth. Without it, long-distance radio communication would be exceedingly difficult and expensive. Although long-range communication via the ground wave is possible in bands 4 and 5, the cost of equipment and antenna systems is high, and the interference effects of natural and manmade noise are damaging. As the frequency is increased, noise levels become lower, and equipment simpler and less expensive.

Up to frequencies of the order of 30 MHz (10 meters), nearly all communication over significant distances is by the sky wave, by ionospheric propagation.

If the air around the Earth were uniform, the component of the wave traveling into the sky (hence the term *sky wave*) would, in a fraction of a second, be tens of thousands of miles out in space. (In 0.10 second, a radio wave travels 18,600 miles, or 30,000 km.)

Fortunately, the blanket of gases surrounding the Earth is not homogeneous, and the ionosphere can return these radio waves to Earth. As the radio wave travels into the ionosphere, its direction of motion and its velocity begin to change. Slowly, as it penetrates more deeply, it continues to bend downward, until finally it leaves the ionosphere, heading back to Earth. The angle at which it leaves the ionosphere is the same as the angle at which it entered (Figure 44). By the time the wave is back in "normal" air, it is traveling with its original velocity. On reaching the Earth's surface, the wave is reflected and "bounces" back toward the ionosphere again. The angle of this bounce or reflection is identical to the angle at which it left the transmitting antenna.

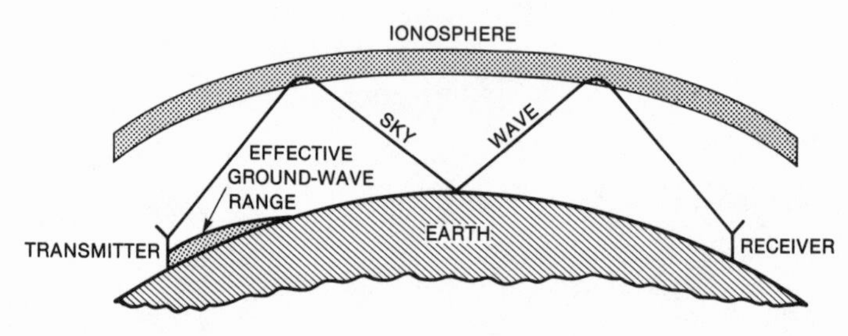

Fig. 44. An example of wave propagation from transmitter to receiver in two hops.

Upon reaching the ionosphere again, the wave is once more bent and is returned to Earth a second time. Each of these excursions is known as a *hop*. Figure 44 shows two such hops.

In band 7, commonly known as the high-frequency portion of the spectrum, the way the sky wave propagates depends on many factors, including time of day, season of the year, and geographical location. During the daylight hours, for example, ionization density in the layers of the ionosphere absorbs frequencies at the lower end of band 7, and the ionosphere propagates primarily the higher ones. At night, the situation is reversed. The higher frequencies penetrate the ionosphere and are lost in outer space, and the lower frequencies are transmitted. This is because radio waves entering the ionosphere are bent in proportion to the ionization density, which is lowest at night. Bending is proportional to frequency as well as to ionization density; the higher frequencies are not bent enough at night to return them to Earth.

The manner in which a radio wave is returned to Earth depends also on the angle with the horizon at which it leaves the Earth on its trip skyward. The smaller the wave angle, the less bending will be required to return it to Earth, and the farther from its transmission point will it be returned. This is pictured in Figure 45.

The angle a radio wave makes with the horizon as it leaves the transmitter is called the *angle of radiation*. If this angle is increased beyond a certain point, called the *critical angle*, for any given frequency, the wave passes through the ionosphere, and is not returned. There is a *critical frequency*, below which a wave beamed straight up will be returned, while a wave of higher frequency continues through the ionosphere and is lost in outer space. This frequency varies greatly according to the time of day, the season of the year, and a number of less important factors. For a wave at some other angle than vertical, signals at frequencies higher than the critical frequency may be returned to Earth. Such waves are returned to Earth at distances depending on the angle of radiation. Signals will be heard near the transmitter; then there is a zone, called the *skip zone*, in which no signal can be heard. The distance between the end of local—ground-wave—coverage and the beginning of the area in which the reflected sky-wave signals are received is the *skip distance*. At any given time, the higher the frequency, the greater the skip distance.

Ionization depends on radiation from the sun. Therefore, ionization density varies from one point on Earth to the next, and from one season to another, because these factors depend on the angle of the sun in the sky. Still another factor influences the condition of the ionosphere, a cyclical

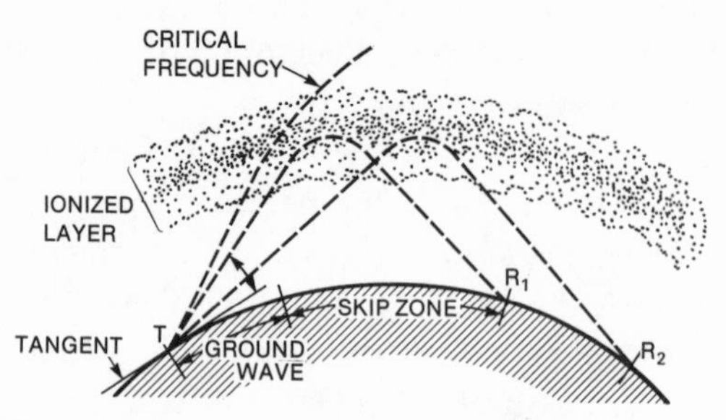

Fig. 45. Effect of the ionosphere in returning a signal of a given frequency, beamed at various angles, to Earth.

one that depends on the number of sunspots on the surface of the sun. Sunspot activity varies over an average eleven-year cycle, going from minimum to maximum in approximately seven years, then returning to minimum in four. No satisfactory explanation for sunspots exists, but they are a source of additional ultraviolet radiation from the sun. So, during years of high sunspot activity, ionization at any given time and location is greater than during a corresponding low year. As a result, the range of frequencies the ionosphere reflects during active sunspot years is greater than during minimum years.

PROPAGATION ABOVE 30 MHz

Propagation characteristics in bands 8 and above change radically from those in bands 4 to 7. Although propagation via the ionosphere does take place during years of maximum sunspot activity, it drops off rapidly as frequencies rise significantly higher than 30 MHz.

There are many forms of radio-wave propagation above 30 MHz over considerable distances. The F_2 layer of the ionosphere sometimes sustains frequencies in band 8 during several hours of the day in years of high sunspot activity. The condition is not considered normal ionospheric propagation.

VHF and UHF signals, bands 8 and 9, are sometimes bent as a result of sharp changes in the refractive index within the troposphere. This effect occurs principally at the boundaries between air masses and frontal systems. Signals sometimes travel far beyond the horizon, to distances of several hundred miles and more, because of such conditions.

Sometimes an effect known as *ducting*, or *trapping*, occurs: two masses of dissimilar air, one lying over the other, act like a duct; the signal, once within this duct, is transmitted in it to distances of hundreds, or even thousands, of miles, depending on the extent of the weather system causing the effect.

Some of the weather factors that affect radio transmission in the VHF and UHF bands are rapid cooling of the Earth after a hot day, rapid heating of the upper air at sunrise during the summer, two dissimilar air masses—such as hot and dry versus cool and moist—coming into contact, and the flow of cool moist air into valleys during the early evening during the summer months. All these factors can produce long-distance tropospheric radio-wave bending. These effects are not reliable and cannot be counted on to occur with any degree of regularity.

SPORADIC-E PROPAGATION

Sporadic-E clouds, or patches, are areas of exceedingly high ionization that form within the E layer of the ionosphere. They reflect radio waves of considerably higher frequency than any of the normal layers of the ionosphere.

The first record of sporadic-E propagation dates back to the mid-thirties. Amateurs working in the supposedly line-of-sight 56-MHz band reported contacts ranging up to 1,200 miles. The observations opened the field of VHF propagation wide, for it was believed that long-distance communication at frequencies much above 30 MHz was impossible.

Sporadic-E regions in the ionosphere cover relatively small areas; the ionized region itself is usually rather thin as ionospheric layers go, and recent measurements indicate that their motion through the ionosphere is swift, ranging up to 300 miles per hour in a westerly direction.

Because the effects of these regions are relatively short-lived and appear in a more or less random fashion (except for specific variations which we shall discuss shortly), they are called *sporadic*. The *E* comes from the region in the ionosphere in which they occur.

Some sporadic-E clouds have been known to reflect frequencies in the 2-meter band (frequencies from 144–168 MHz), although such occurrences are rare. In general, sporadic-E propagation is most prevalent in the bands above 6 meters.

Sporadic-E density can be great enough to present a highly efficient reflecting surface to the incident wave. This mirror effect occurs with a minimum of absorption, and results in reflecting an extremely strong signal to Earth. The phenomenon often occurs on the longer wavelengths of 15 and 20 meters (20 and 15 MHz) at times when the regular layers of the ionosphere will support these frequencies also. However, sporadic-E reflected signals are absorbed so much less than regular F-layer reflected signals that the effect at these times is to override all other signals in these bands (see Figure 46).

Propagation by way of sporadic-E "clouds" is, in most cases, limited to distances of 1,500 miles or less, because the limited geographical area covered by a patch of sporadic-E usually rules out any possibility of multihop propagation. But there have been cases where either simultaneous occurrence of sporadic-E patches over widely separated areas, or a combination of sporadic-E clouds and regular F_2-layer reflection, made it possible to communicate over a circuit when it would have been impossible to do so by way of the regular F_2 layer alone.

Sporadic-E propagation, because its range is generally limited to relatively short one-hop paths, is also called *short skip*, although the use of the term *short skip* does not necessarily mean that propagation is always via a patch of sporadic-E ionization.

Although sporadic-E propagation has been known for forty-five years, a satisfactory scientific explanation for all the effects attributed to sporadic-E patches is yet to be found, and evidence indicates that there may actually be more than one kind of sporadic-E. For example, sporadic-E occurs most frequently in midlatitude during summer daylight hours. This would indicate a connection with ultraviolet radiation.

The most widely accepted theory holds that sporadic-E clouds are formed by the shearing forces of winds in the upper atmosphere, which

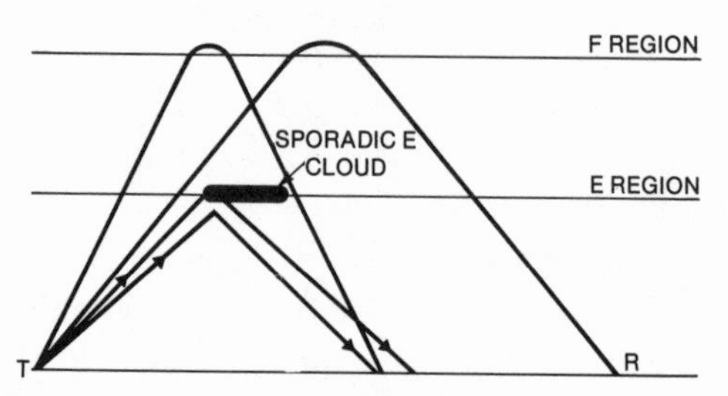

Fig. 46. Interesting effects can occur when a signal is reflected from both the E and the F layers.

concentrate areas of high-velocity ionized gases in a very narrow region; the effects of shearing forces would account both for the high ionization density of sporadic-E clouds and their high speed of travel.

Variations in sporadic-E activity have also been observed with geomagnetic latitude. In the auroral zones, for example, sporadic-E occurs most frequently and is particularly prevalent during ionospheric storms. (To complicate the situation further, there is also peak sporadic-E activity in equatorial regions.)

SCATTER PROPAGATION

In 1932, Marconi predicted that VHF and UHF radio communication at distances considerably beyond the line-of-sight was a future likelihood. At the time of the prediction, he had already achieved some success in sending a UHF signal 165 miles in the 500-MHz range. During the years immediately following, exploration of the VHF and UHF ranges really began.

By the end of World War II a great deal of information about the propagation of radio waves on frequencies above 30 MHz had become available. This was partly due to the development of radar, with its very high radiated powers and high-gain receiving systems operating in the VHF and UHF ranges.

As use of frequencies above 30 MHz increased, an increasing number of reports of "anomalous" propagation began to come in. Many of these told of strong signals being received for relatively long periods of time over paths and in frequency ranges that were thought to be impossible to use.

Circuit analyses indicated that signals were being propagated through the troposphere, the region of the Earth's inner atmosphere closest to the Earth, in which are most of the cloud formations and weather phenomena on the planet. No theory satisfactorily explains the propagation of VHF and UHF signals for long distances through the troposphere. It is generally believed to be connected with scattering from areas of turbulence associated with the movement of weather fronts and with temperature gradients. VHF and UHF signals encountering these turbulent areas are scattered in different directions. Although this scattering results in large losses of signal strength in any given direction, with high-gain transmitting and receiving systems reliable communication is possible over paths that cannot be served by any other known communication technique.

Although tropospheric scatter has been observed to occur both in the

VHF and UHF portions of the spectrum, it has been found that results are optimum in the range between 400 and 2,000 MHz. Scatter links use highly directional transmitting and receiving antennas, as well as high-effective radiated powers, in the order of thousands of kilowatts.

Tropospherically scattered signals have two fading components, one rapid, the other slow. There are seasonal variations in the signal levels; the strongest signals are observed during the summer. There are also meteorological variations, with best signal levels when the air of the troposphere is very warm or very dry.

Although the frequencies most successful in tropospheric-scatter propagation are in the range of 400 MHz upward, it has been found that the efficiency of the circuit decreases as the frequency increases. On the other hand, efficiency does not appear to decrease with increased distance at the same frequency. Thus, at a given frequency, there is no appreciable difference in efficiency on circuits from one hundred to several hundred miles distant. Most scatter links are not very reliable for communications beyond 500 miles, with most links much shorter. Experiments have shown that fair reliability can be attained in some cases on circuits up to 750 or more miles in length.

During recent years a great many significant advances, especially by the military, have been made in tropospheric scatter propagation. A number of such links relay wide-band telephone signals up to several hundred miles with amazingly high reliability. Because of the propagation losses in tropospheric scatter, fantastically high powers are required, and it is only this reliability that makes the technique practical. Shortwave transmission would do the same work at probably a millionth of the power, but cannot be depended on for absolute reliability (blackouts of three days' duration, because of "magnetic storms," are not unknown). Satellite communications may make scatter communications obsolete in the near future.

IONOSPHERIC SCATTER—METEORS

The mechanism of ionospheric scattering is believed to be similar to that of tropospheric scattering. Sporadic-E patches move westward at velocities up to 300 miles per hour. This motion is believed to be due to ionospheric "winds" which blow like winds here on the surface of the Earth, but with much greater intensity. These ionospheric winds cause considerable turbulence in the ionosphere. As a result there are very rapid fluctuations of refractive index in certain regions.

It is generally believed that ionospheric scatter occurs because of these random fluctuations of the ionospheric refractive index. The total energy received at the antenna is then made up of all the different contributions from the many regions that contributed to the scattering of the signal.

Yet another theory attempts to explain ionospheric scatter. It accounts for scatter solely on the basis of ionized meteor trails, which are formed continually in the regions of the E layer. Two kinds of meteors arrive in the Earth's atmosphere. One of these is the regular shower type, which occurs at definite intervals. The Perseids, which appear every August, are an example of these showers. There are also those that arrive daily, at random from random directions. It is estimated that hundreds of millions of these arrive in the earth's atmosphere on any given day.

A meteor, because of its very high travel velocity, produces a trail of ionized gases. Radio signals can be reflected from that ionized trail. The heights of these reflected meteor-trail signals have been calculated, and the results indicate that they are around 110 km above the earth, which is the vicinity of the E region (Figure 47).

This proximity to the E region of the ionosphere has led some scientists to speculate that sporadic-E activity was at least partly attributable to ionization by meteors. This theory does not explain summertime peaks in sporadic-E activity.

Both the turbulence and meteoric ionization theories predict that the useful range of frequencies would run approximately from 25 to 60 MHz, after which the strength of the signal would drop off rapidly. A great many experiments conducted during recent years tend to confirm this.

Before ionospheric-scatter propagation was discovered, it was be-

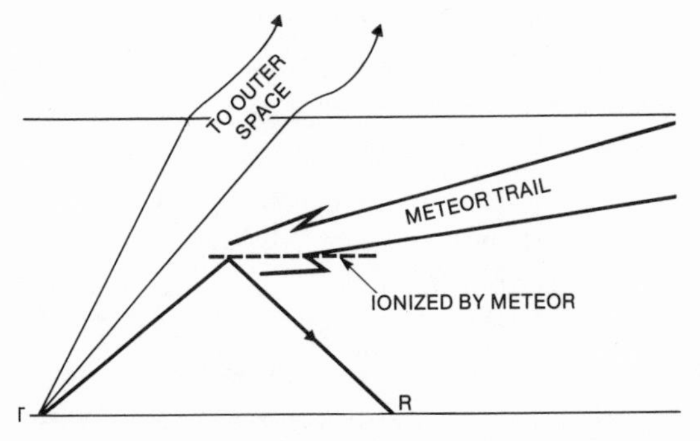

Fig. 47. Transmission by meteor-trail reflection.

lieved that as the frequency used for transmission was raised above the optimum working frequency, the delivered signal intensity would drop off, until eventually it would disappear completely. Several experiments conducted in the United States recently have shown that this is not the case; that actually the signal reaches a certain minimum value. This minimum level is attributable to ionospheric scatter.

As in tropospheric-scatter propagation, highly directive transmitting and receiving antennas are required to carry on reliable service, with effective radiated powers of the order of hundreds of kilowatts usually being necessary to maintain adequate signal-to-noise ratios. Scatter circuits have been used effectively for some time by the military. Highly reliable telegraph circuits have been maintained; the optimum distance for this kind of service is between 600 and 1,200 miles. Another highly important aspect of ionospheric-scatter propagation is that signal levels are generally maintained or perhaps even enchanced during periods of ionospheric storms.

AURORAL PROPAGATION

During the 1930s, amateurs working on the 5-meter (60 MHz) band found that during periods of severely disturbed radio conditions they could communicate over distances considerably greater than normal by beaming signals in a northerly direction, rather than toward each other.

Further investigation showed that during the night this phenomenon was generally associated with a visible display of aurora borealis. The conclusion was thus drawn that auroral displays also occurred during the daylight hours, even though they could not be seen. Although little was known in those early days about the aurora, the numerous observations made by radio amateurs throughout the world have helped immeasurably in furthering our knowledge about it.

Dr. Carl W. Gartlein of Cornell University, who had long been one of the outstanding auroral researchers, made considerable use of the data submitted by radio amateurs throughout the United States, Canada, and Alaska. After World War II, auroral propagation was extended into the 2-meter (144 MHz) band, where contacts in recent years have occurred at least as frequently as at 6 meters, or 50 MHz. To date, the highest known frequency to be "bounced off the aurora" is in the vicinity of 230 MHz. In general, the 10-meter (28 MHz) band represents the limit of effective auroral propagation. Lower frequencies do not propagate.

The characteristics of aurora-propagated signals vary. On the lower frequencies, there is a characteristic auroral flutter fade, but as the frequency is raised, the note of a continuous-wave signal begins to resemble a hiss somewhat like the sound of steam escaping from a radiator.

Since the fading component frequency ranges from 100 to 2,000 hertz, radio telephony is generally chopped up and rendered unintelligible. Therefore, a slow continuous-wave signal has a much greater chance of success.

The importance to amateurs of auroral propagation has been far-reaching. It has enabled stations up to 800 miles apart, lying in an east–west direction, to communicate by beaming their signals to the north. Since auroral phenomena are generally associated with severe ionospheric storms, during which normal shortwave radio contact is either impossible or very difficult, their importance cannot be minimized.

In general, auroral displays are most common during the spring and fall and occur more frequently during periods of maximum sunspot activity than at other times, with peak activity usually occurring from two to three years after sunspot maxima. The next period of maximum auroral propagation should therefore be in 1980 and 1981.

Recent experiments have shown that auroral reflection takes place from a region behind the visible aurora and is best when the incident radio waves are beamed in a direction perpendicular to it. It would therefore seem advisable to keep the vertical radiation angle as low as possible so that the effective reflecting surface of the aurora will be at right angles to the incident wave.

Although several theories try to explain auroral propagation, none of them is wholly satisfactory. It is generally believed that the incident radio signals are scattered from primary and secondary ions resulting from the bombardment of the Earth's atmosphere by corpuscles coming from the sun.

NOISE

Successful radio communication, whether in the shortwave portion of the spectrum or on higher frequencies, depends to a great extent on the levels of background noise at the receiving station. Broadly speaking, four major categories of noise must be taken into account. These are cosmic noise, atmospheric noise, receiver noise, and manmade noise.

Cosmic Noise. This reaches us from extraterrestrial sources. It is generated, among other places, in the sun, in "radio stars," and in interstellar gas clouds. It predominates at frequencies from 15 MHz up, becoming of secondary concern above frequencies of 1GHz, where noise generated in the receiver itself is most important.

Atmospheric Noise. This type occurs primarily in the troposphere and is due mostly to thunderstorm activity. It is important only on frequencies below 100 MHz, and becomes troublesome in bands 5 and 6.

Receiver Noise. This is produced mainly in the receiving antenna, the transmission line, and the front end of the receiver. It is thermal in origin, being caused by random molecular motion. It increases with temperature and the bandwidth of the receiver.

Manmade Noise. When this occurs, it is generally stronger than all the other noise types. It is generated by electric motors, diathermy machines, auto ignition, and many other sources. The wisest course of action is to avoid locating a receiving site in an area of high manmade noise if it is at all possible. Fortunately, manmade noise occurs relatively infrequently at frequencies above 500 MHz. Since this type of noise generally emanates from a particular location, one limited method of combating its effects is to use a highly directional antenna system.

14

Rise of the Shortwaves

The radio amateurs' remarkable exploitation of the "below-200-meter" bands during the early 1920s shook the commercial companies. RCA had already allocated funds for a worldwide communications network operating on low frequencies with Alexanderson alternators. Other companies had a similar "set" in favor of the long waves.

The Marconi Company had also emphasized operations on the longer waves. Marconi's assistants, C.S. Franklin and H.J. Round, conducted a number of experiments in the shortwave region of the spectrum after World War I, but it was not until after the news of the amateurs' exploits that the Marconi Company undertook shortwave operation in earnest.

Low frequencies were indeed reliable during the winter months. But with warmer weather, atmospheric noise levels because of thunderstorm activity rose rapidly and reliability decreased sharply. Shortwaves, on the other hand, are reliable throughout the year, though on different frequencies at different times, and are much less vulnerable to atmospheric noise, although ionospheric "blackouts" disrupt shortwave communication periodically.

The blackouts are caused by solar flares, believed to be due to sudden changes in the magnetic fields associated with sunspots. They occur suddenly and create enormous X-ray and ultraviolet radiation, which reaches the Earth in about eight minutes. Solar flares also eject vast quantities of matter, including protons and electrons. These travel at relatively low speeds, reaching the Earth in eighteen to thirty-six hours after the flare. The radiation produced by the flare causes abnormally high absorption in the D region of the ionosphere, disturbing shortwave communication over portions of the daylight hemisphere of the Earth. These disturbances occur quickly and rarely continue longer than an hour.

The ionospheric storm, or blackout, is caused by the particles from the flare. Unlike the Sudden Ionospheric Disturbance (SID) caused by the radiation, ionospheric storms occur from one to three days after the flare; they are generally worldwide, occurring in the day and night hemispheres simultaneously, and continue for several days. They are associated with magnetic storms, which are characterized by wide variations in the Earth's magnetic field.

Spurred by continuous reports of amateur successes, RCA had completely abandoned its plans for a worldwide long-wave network by the mid-1920s, and began to install shortwave transmitters in the United States and abroad. The higher frequencies were found not only more reliable but more economical to operate. The shorter wavelengths required smaller antenna systems. They took up less space and were cheaper to construct. Directional transmission became possible with these smaller antennas. The directional antenna systems made it possible to focus transmitted energy on a particular reception point, keeping radiation in other directions to a minimum, thus delivering more of the transmitter's power to the distant receiver. At the same time, receiving antennas could also be made directional. When aimed at the transmitter, unwanted signals from other directions were attenuated, thus improving the effectiveness of the system even more.

The simplest directional antenna radiation pattern is shown in Figure 48. It is produced by a single antenna element half a wavelength long. (At a wavelength of 15 meters, the antenna element is approximately 7.5 meters long.) The pattern gives relative values in the horizontal plane.

Engineers soon learned that directivity could be increased by using additional elements, either in front of (director) or behind (reflector) the radiating element. Elements longer than the radiator tend to reflect the wave; elements shorter than the radiator tend to direct it. The system in Figure 49 has both a director and a reflector. Such a system directs the radiation very sharply, as shown in Figure 50.

By 1924 vacuum tubes that could operate at high power and remain in service continuously at high frequencies had been developed, and RCA put the first commercial shortwave transmitter into service, operating on a wavelength of 103 meters (about 3 MHz) to carry message traffic. By 1926 commercial shortwaves services on frequencies up to 20 MHz (15 meters) were in operation in the United States, Europe, and Latin America.

The ability of the short waves to span enormous distances with high

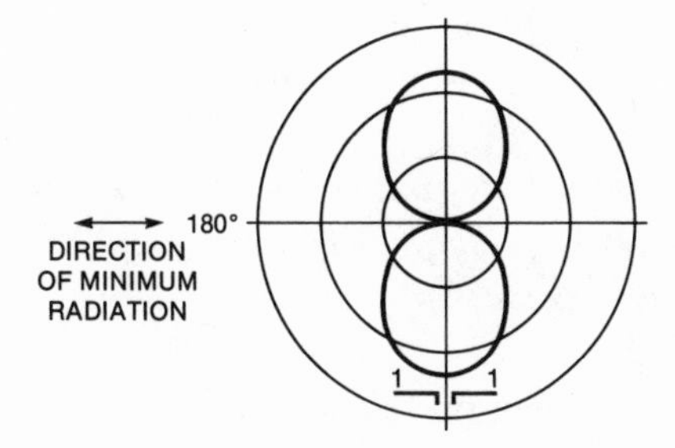

Fig. 48. The horizontal radiation pattern of a simple single-wire (dipole) antenna, half a wavelength long.

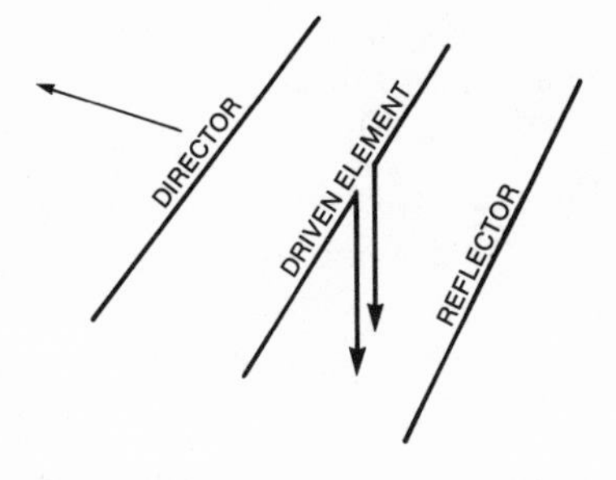

Fig. 49. A three-element antenna system, with one director, one reflector, and a driven element.

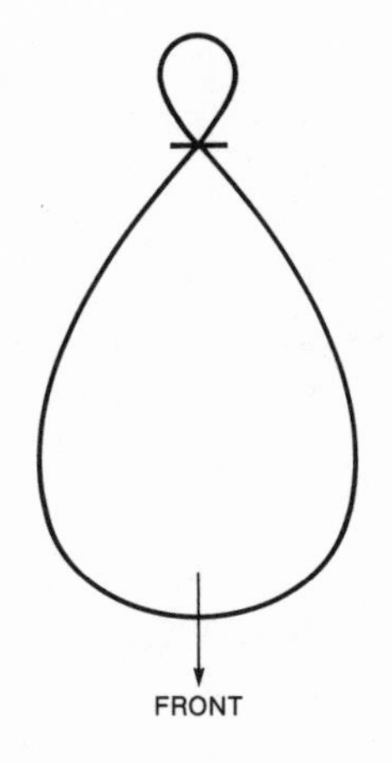

Fig. 50. Horizontal radiation pattern of the antenna system shown in Fig. 49.

reliability and at relatively low cost made them ideal for use in intercontinental radiotelephone and telegraph services. It was not long before it was realized that these same short waves could be used to carry voice transmissions to foreign countries, either by broadcasting them so that they could be picked up in homes with shortwave receivers, or beaming them for pickup and relay by other broadcasting organizations. This led to an exchange of program material, relayed via shortwave, between national and international networks.

The shortwave spectrum became an important subject of discussion at the international conference of 1927. Major changes had taken place in the fifteen years since 1912:

• The vacuum tube had been developed and improved to a point at which it could operate at high power and high frequencies with excellent reliability. As a result, voice transmission, particularly broadcast services, had become highly successful.

• The range of useful frequencies had been pushed far into the shortwave region. But no internationally agreed-on allocations existed.

• Radio communication with aircraft had been introduced and was being used routinely.

• Commercial communications companies were increasingly using the short waves to carry teletype and telephone traffic.

Representatives from eighty countries and more than sixty private commercial carriers were present at the Washington conference of 1927, and the decisions taken were far-reaching. Herbert Hoover, the Secretary of Commerce, was elected conference chairman. One of the significant accomplishments of the conference was to impose restrictions on spark transmitters. With spectrum use increasing rapidly, the broad-band sparks were taking up valuable space, as well as causing interference to other services. A motion to outlaw sparks entirely was narrowly defeated—maritime interests contended that sparks were simple to operate and that the broad bands on which they operated were an advantage for distress calls. In spite of their stand, the maritime interests gained only a delay, and the conference voted to put an embargo on the installation of any new spark sets, and to ban these entirely on ships by January 1, 1930.

Another major accomplishment of the conference was to allocate the frequency spectrum that was then in use (frequencies from 10 kHz to 60 MHz). The history-making conference allocated frequencies to broadcasting services, the aeronautical services, and radio amateurs, who for the first time were given official international recognition. The amateurs were assigned some 7,500 kHz of spectrum space as opposed to 12,000 kHz assigned by the United States, and although they lost some 40 percent of their American allocation, they had been provided for in an international treaty agreement.

Directional antenna designs continued to advance. Additional elements and configurations of elements led to higher gain, increased directivity, and stronger signals, and experiments in international broadcasting on shortwave were undertaken. By 1932, the BBC had inaugurated a world service on shortwave, which it called Empire broadcasting. The broadcasting transmitter bandwidth was 8 kHz, making it possible to transmit relatively undistorted voice, and music with fairly good quality.

From the late 1920s, several American manufacturers, such as General Electric, RCA, and Westinghouse, had been operating their own experimental shortwave broadcast stations. The success of domestic broadcasting led to pilot programs to determine whether international broadcasting carried out by private interests could make profits. During World War II the United States government leased these facilities to carry programs overseas, both to entertain our troops abroad and to beam programs to the neutrals, the Allies, and the Axis powers.

At the end of the war the United States government entered the field of international broadcasting with the formation of the Voice of America, whose mission it was to bring the story of the United States and its policies to the world. At first, transmitters owned by GE, RCA, and Westinghouse were used. The private leasing of facilities was gradually discontinued, and today the Voice of America owns all its United States transmitting facilities.

Another major International Radiotelegraph Conference was held in Madrid in 1932. It defined the term *telecommunication* for the first time, as "any telegraph or telephone communication of signs, signals, writings, images, and sound of any nature, by wire, radio, or other system or process of electric or visual (semaphore) signalling."

This compares with the current (Radio Regulations, Geneva, 1968) definition of the term: "Any transmission, emission or reception of signs, signals, writing, images and sounds or intelligence of any nature by wire, radio, visual, or other electromagnetic systems."

The Madrid conference merged all existing telecommunication services into a single entity, the International Telecommunication Union, and drafted a set of regulations, termed the Convention. Plenipotentiaries from all member countries of the Union meet periodically to bring the Convention up to date.

During the 1930s shortwave developed along a number of fronts. Diversity reception was developed. Observers had noted that shortwave signals could fade out at one location and at the same instant be strong on a receiver only a few hundred yards away. By using several antennas, spaced 1,000 feet apart, for example (the optimum distance depends on a number of factors, including frequency, distance between transmitter and receiver, antenna heights, etc.), the effects of fading could be minimized. The antennas are connected to separate receivers, the outputs of which are combined, giving a more constant, fade-free signal.

In another major refinement, oscillators were designed in which a

Fig. 51. David Sarnoff, left, and Guglielmo Marconi, at RCA's Rocky Point, New York, transmitting station in 1933. *Courtesy of RCA.*

crystal precisely controlled the frequency. The output of such an oscillator is extremely stable but usually at a power level low enough to require amplification through several stages of vacuum tubes to produce sufficient power to control a large output tube. It was also possible to use these amplifier stages to multiply the oscillator frequency without sacrificing stability inherent in the crystal-controlled stage. This was yet another advantage, because oscillators tend to become more complex and difficult to operate to achieve desired stability, as frequency increases. Thus, high-powered, high-frequency transmitters became practical.

In the 1930s, giant transmitting and receiving sites were constructed. These nerve centers for commercial high-frequency communications carriers had scores of high-powered shortwave transmitters and receivers; fields of antennas so extensive they were termed antenna farms covered acres of land.

By 1938, the demand for channels in the shortwave spectrum was extensive. The Cairo Conference held that year had to grapple with this and other problems. For the first time, frequencies were assigned to international aviation routes and to services operating in the tropics; further changes in amateur allocations were made, as well as additional assignments to the broadcasting service at the upper end of the shortwave portion of the spectrum.

The use of spark transmitters, which were still hanging on, was further restricted, confined now to only three frequencies: 375, 425, and 500 kHz. A portion of the spectrum in the VHF range was allocated to various air, radio-sounding, mobile, and some fixed services, and a part of the VHF region was allocated to an infant industry which would not begin to grow until after the imminent World War II, but which would then become one of the great forces in the history of communication—television.

For even while shortwave techniques were being developed and improved, and communication in the 3 to 30 MHz region was becoming technically and economically feasible, a new and even more significant era in telecommunications was dawning. In the late 1920s and 1930s, great technical advances were being made in other fields. Radio detection and ranging (radar) and television would soon change the world in which the spark transmitter had only thirty-five years before begun to crash out messages at an agonizingly slow rate.

15

Television and FM

One of the most remarkable achievements of this—or any—century has been television—instantaneous transmission of visual images over great distances. Throughout his history, man has dreamed beyond his capabilities. Progress is the realization of those dreams. The concept of man in flight dates back to Icarus and the mythology of the Greeks, long before science made flight possible. Similarly, the dream of transmitting pictures and associated sounds through space dates back at least to the early 1800s, one hundred years before the engineering and scientific tools that would make it possible.

Various means of transmitting pictures to distant points had been proposed. Among them was the idea that the images to be transmitted would have to be broken down and sent bit by bit, rather than being sent as complete pictures. In 1884, scientist Paul Nipkow obtained a German patent for doing just that. It consisted of a rotating disc with a series of apertures arranged spirally around it, as shown in Figure 52. An image is projected onto an area near the edge of the disc. As the disc rotates, the apertures move across the image in succession. As each aperture passes over part of the image, the light passing through it varies in proportion to the brightness of the image at each point. Successive apertures scan the image at points slightly below the one above, and each time a succession of brightness impulses passes through the aperture. As the disc continues to rotate, each of the apertures traces out different values of light intensity corresponding to lines running horizontally across the image. By the time the disc has rotated once, a series of lines has been traced out across the image, from top to bottom. Nipkow reasoned that if a photosensitive material was used to convert the different values of light intensity passing

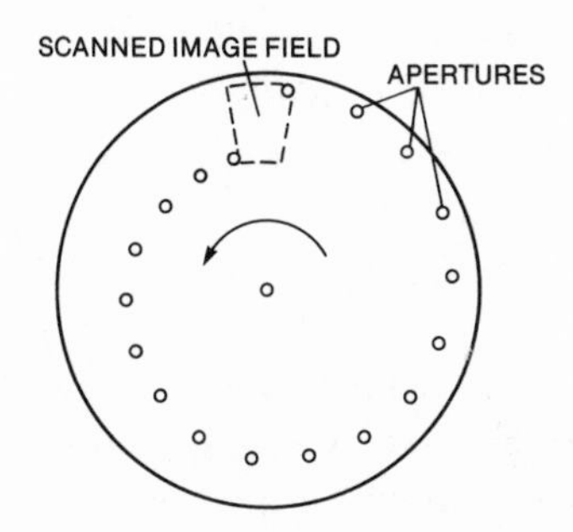

Fig. 52. A Nipkow disc, the original television scanning apparatus.

through the apertures into electrical impulses, and if these were transmitted to a receiving apparatus, turned back to light impulses, and projected through a synchronized rotating disc with an identical number of apertures, an image could be produced that would be the exact duplicate of the one projected on the transmitting disc.

If the receiving system operated precisely in synchronism with the transmitter, it would reassemble this series of brightness values, and if the assembly was fast enough, the viewer's persistence of vision would give the illusion of a complete image. (Successive images blend together if projected onto the eye within 0.1 second, giving the impression of a continuous image. The illusion of movement in motion pictures depends on this persistence of vision. Actually, individual frames are projected on the screen at the rate of about 24 a second.)

Unfortunately, the only light-sensitive material known to Nipkow was selenium. It can convert light into electrical impulses, but does not operate fast enough to produce a picture of any quality or definition. A second problem is that the signals produced by the light beams passing through the aperture are very weak, and in Nipkow's time there were no amplifying devices that could boost those weak signals to the point that when converted to light, they would be bright enough to project a recognizable image on a screen.

So the dream of television had to wait for technological developments to catch up to it. Nipkow, who had conceived the system, was never able to reconstruct his transmitted pictures at the receiving disc because selenium was too sluggish, and because the signals were too weak.

Five years after Nipkow's disc, a German, Lazare Weiller, conceived another scanning system. It used a drum on which small mirrors were mounted. Each mirror did the same work as a hole in Nipkow's disc, and the beams from the mirrors were much stronger than those through the holes in the earlier disc. Professor Weiller, too, was ahead of his time, and although he was one of the pioneers of television, he could not produce an operational system.

In the meantime, other developments, seemingly unrelated, but highly significant in the evolution of television, were taking place. Notably, Karl Ferdinand Braun in Germany was developing the cathode-ray oscilloscope. Carrying forward the work of Sir J.J. Thomson, Sir William Crookes, and others, Braun's tube could display changes in the strength of a stream of electrons visually on the face of the tube.

A simple cathode-ray tube is shown in Figure 53. The cathode C is a filament which, when heated, emits electrons. The anode A is a disc with a small hole in its center. The plates P_1 and P_2 are arranged in pairs at right angles to each other. When the filament is heated and a voltage applied between it and the anode, an electron stream (a cathode ray) passes through the hole and continues in a straight line through the plates until it strikes the screen S, which is coated with a fluorescent material; the electron beam makes a tiny luminous spot where it strikes the screen.

When the plates are charged, the electric field between them deflects electrons passing through; the plates P_1 deflect the beam either upward or downward, while P_2 deflects the stream to the right or left. If the voltage applied to the plates is varied, the electron stream moves, tracing out a glowing line, which depends on how the voltages on the plates are varied. The brightness, or intensity, of the beam coming through the anode depends on the potential applied to it. It becomes immediately apparent that if the potential applied to the anode is varied with the brightness of the light passing through a Nipkow disc, then a means of reproducing an image has been found. (The modern television tube, Figure 54, deflects the beam

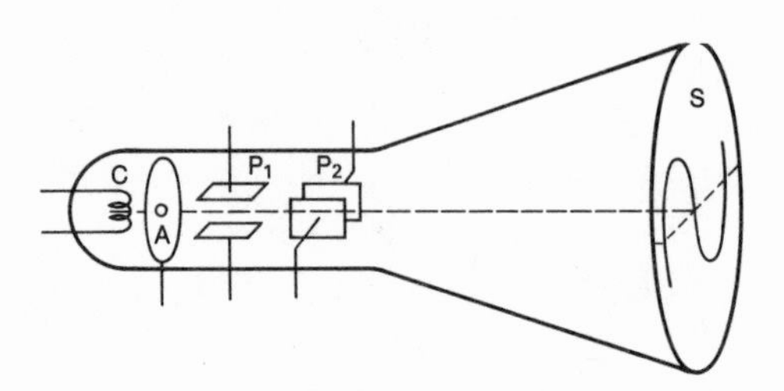

Fig. 53. A simple cathode-ray oscilloscope. *From* Physics, *Hausmann and Slack.*

with magnetic coils around the neck of the tube—there is some evidence that Braun's original cathode-ray tube also used magnetic deflection. The brightness of the beam is controlled by varying the voltage on a "grid" element, rather than varying the anode voltage.)

In 1907 the Russian scientist Boris Rosing patented a system that could transmit and reconstruct simple geometrical shapes by using a mirror drum scanner to transmit images, and a Braun oscilloscope at the receiving end to assemble the picture (Figure 55).

The following year, the Scottish electrical engineer A.A. Campbell Swinton (the man who introduced Guglielmo Marconi to William Preece of the British Post Office) made a remarkable proposal in a letter to the scientific journal *Nature.* He suggested that a cathode-ray tube be used both at the transmitting and receiving ends of the circuit. Amplifying his ideas before the Roentgen Society of London in 1911, he proposed a transmitter

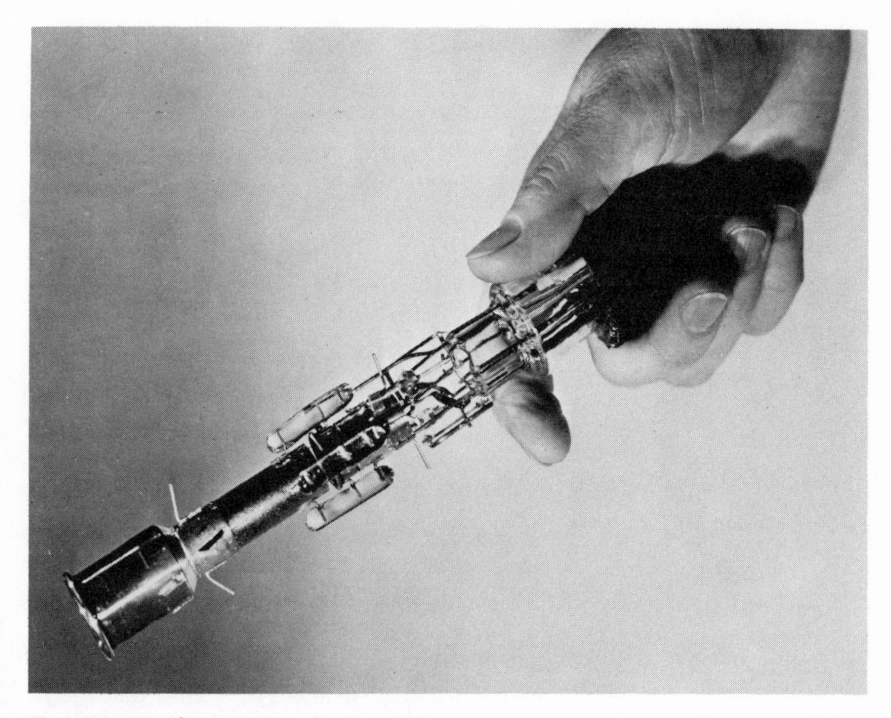

Fig. 54. An electron gun, heart of the television picture tube. *Courtesy of the General Telephone and Electronics Corp.*

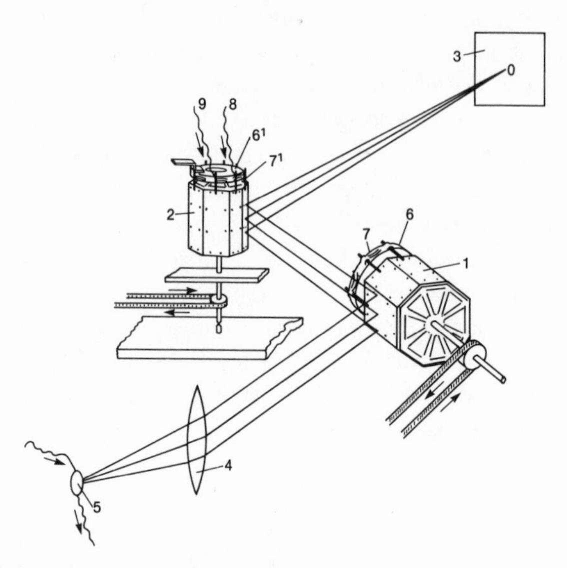

Fig. 55. Rosing's 1907 television patent.

composed of a mosaic screen of photoelectric material upon which the image to be sent would be focused. A beam of electrons emitted at the cathode would sweep the image, line by line. The electrons sweeping the image would cause electrical discharges whose intensity was proportional to the brightness of the image at each point of impingement. The picture would be reassembled at the other end by a cathode-ray tube that emitted electrons in proportion to the strength of the incoming impulses and in synchronization with the scanner at the transmitter.

Up to this point, no one had visualized sending pictures through space with wireless waves. Even Campbell Swinton's idea, which describes modern television fairly accurately, visualized a closed-circuit system, with transmitter and receiver connected by wires. By 1914, with the first production of continuous radio waves generated by vacuum tubes, the idea of transmitting television pictures through space without wires was beginning to seem feasible. But World War I interrupted further research, and it wasn't until the mid-1920s that further progress was made.

In England, the Scottish inventor John Logie Baird developed the first operational television system, using a Nipkow-type disc, a potassium-filled photoelectric cell, vacuum-tube amplifiers, and a neon light bulb at the receiving end. The pictures Baird transmitted consisted of thirty lines per picture frame, and five pictures per second. The pictures were crude, and flickered badly, but it was the first complete television system to be demonstrated, and the British Broadcasting Corporation used the Baird system for experimental transmissions beginning in 1929 and continuing until 1935.

Other mechanical systems were being developed at about the same time, using either the Nipkow disc or the Weiller drum. (One was actually offered commercially, by Francis Jenkins and Lee de Forest.) All of them had the same limitations: an adequate, reasonably clear-definition picture required a minimum of 300 lines per frame. The mechanical disc and drum systems being developed could not operate much above 200 lines per picture frame. Even so renowned a scientist as E.F.W. Alexanderson, working for the GE Company in Schenectady, was unable—with a high-speed rotating cylindrical drum—to develop an image with lines enough to produce a clear, highly defined picture.

Fig. 56. An experimental system that didn't catch on (yet). The screen consists of thousands of tiny lamps. Wires at the rear carry current in accordance with the signals received by a special television set, thus producing a display like that of a kinescope on a larger scale. *Courtesy of General Electric.*

Researchers working on television decided that if it was at all possible to produce a clear picture, it would have to be done electronically. Only two, both in the United States, had taken this approach to the problem. One of these, Vladimir K. Zworykin, had worked with Rosing in Europe and had been impressed by his approach. Zworykin came ultimately to the United States, where he joined the staff of the Westinghouse Electric Company. By 1923 he had filed his first patent on an electronic television tube, but much work still remained to be done on it, and Samuel Kintner, in charge of research at Westinghouse, suggested to him that he concentrate on developing improved photoelectric cells. These, Kintner foresaw, would be instrumental in developing an effective electronic television system.

Although he did as Kintner suggested, Zworykin continued to work on a television tube and by 1928 had developed one that was to be the forerunner of all future electronic television transmitting or "camera" tubes. He called it the iconoscope from the Greek words *eikon*, meaning image, and *skopein*, to watch. The iconoscope is shaped like a dipper, as shown in Figure 57. The image to be transmitted is focused through a lens on a mo-

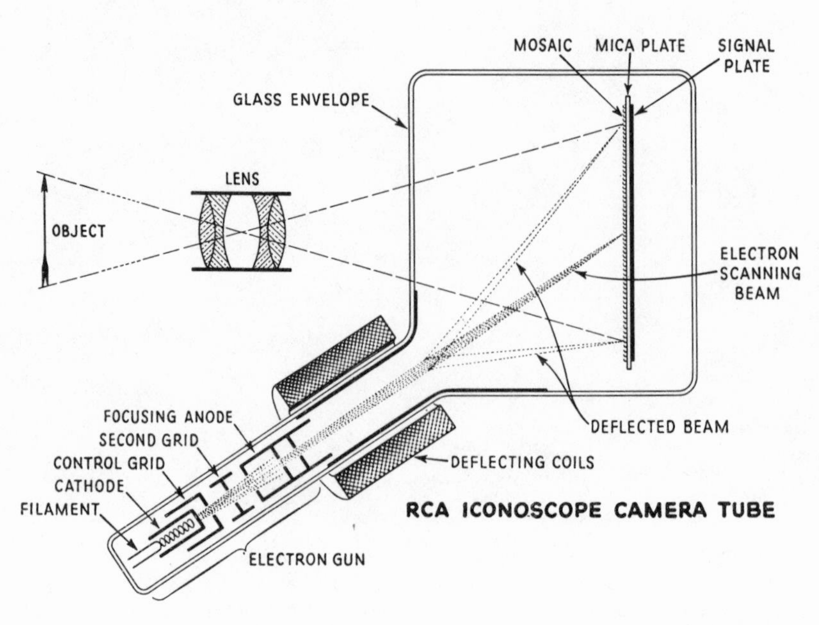

Fig. 57. An iconoscope, the first television camera tube. *Courtesy of RCA.*

saic screen at the "bottom" of the dipper. The mosaic consists of hundreds of thousands of minute globules of silver, treated with cesium vapor and oxygen to produce a coating highly sensitive to light. The treated silver globules, when illuminated with an image, release electrons, making the net charge on the globules positive. The amount of positive charge on the silver depends on the intensity of the light falling on it.

The entire mosaic was made on a thin sheet of mica about 3.5 inches high by 4.75 inches wide. The mica is backed by a metal *signal plate*, which is connected to the control grid of the first tube in the camera preamplifier. The silver globules are insulated from each other, and each forms a small capacitor with the portion of the signal plate directly behind it. Light from the scene to be televised is projected onto the mosaic through a lens system, so the mosaic becomes a screen carrying the image being televised.

An electron gun is located at an angle to the mosaic surface, in the "handle" of the dipper, to keep it out of the way of the scene being projected onto the mosaic. The gun bombards the mosaic with a beam of electrons, first passing through two sets of deflecting coils, which cause it to "scan" the mosaic. The horizontal scanning is extremely rapid, nearly 10,000 feet a second. After each horizontal scan, the second set of coils causes the electron beam to move vertically downward, for the next scan line. The net result is a beam of electrons that scans the image from left to right with a series of horizontal lines close to each other—about 490 lines in the nominal "525-line" TV. The scanning—which Nipkow and Weiller had done mechanically—was thus done electrically by Zworykin.

When the beam of electrons strikes a globule of silver, it supplies the electrons lost because of the light striking that globule. This causes an equal number of electrons attracted as a "bound charge" to the part of the signal plate directly behind the globule to be released. Since the signal plate feeds into the camera preamplifier, a voltage that varies continuously with the amount of light on the globules scanned is applied to the amplifier, causing a signal that varies in intensity with the amount of light along each scan line. Thus an electrical image is built up from the optical one.

The iconoscope was important because it was more sensitive than mechanical (and other electronic) scanning methods. Since the light is projected continuously on the whole mosaic, the globules become more and more positive from the time the beam leaves a globule until it returns in the next scan. This is called a *storage effect*—by the time the beam has returned to a particular spot on the mosaic the charge at that point has

Fig. 58A. Vladimir K. Zworykin demonstrates his cathode-ray television system, developed between 1923 and 1929. *Courtesy of Westinghouse Broadcasting Co.*

increased considerably. As a result, the iconoscope is considerably more sensitive than a similar tube that could not store charge would be.

Once Zworykin had perfected the iconoscope, much work remained to be done in both transmitting and receiving the image. Because of the great speed of scanning, extremely precise methods were needed to synchronize the scanning at transmitter and receiver, to assemble the picture without distortion. In 1930, RCA took over from GE and Westinghouse much of their research in radio and electronics, and Zworykin went to work for RCA.

In April 1923 David Sarnoff wrote in a report to the RCA board of directors:

> I believe that television, which is the technical name for seeing as well as hearing by radio, will come to pass in due course. . . . It may be that every broadcast receiver for home use in the future will also be equipped with a television adjunct by which the instrument will make it possible for those at home to see as well as hear what is going on at the broadcast station.

Sarnoff gave Zworykin his enthusiastic support, supplying him with all the money and personnel he required. By 1932, Zworykin had sixty people working for him as the pressure to develop a commercial electronic television system mounted. Meanwhile, the BBC, the Germans, and the Dutch were experimenting with television broadcasts.

RCA could have marketed television receiving sets by 1933. They would have cost more than five hundred dollars and the quality of the picture would have been poor (at least by today's standards). It was luckily decided, in spite of the apparent progress being made in Europe, to defer commercial operation until a less expensive, higher quality system could be

Fig. 58B. First intercity television transmission, April 1927. Herbert Hoover, then Secretary of Commerce, is televised for transmission via telephone wires from Washington to New York. *Courtesy of Bell Telephone Laboratories.*

developed. Within two years RCA had produced a 343-line picture with a repetition rate of 30 per second. Sarnoff announced plans to transmit pictures experimentally from the top of the Empire State Building, and by 1936 pictures had been transmitted forty-five miles from the new site.

While Zworykin was developing storage-type television tubes, another scientist, working along other lines, developed a second electronic method of picture transmission. Philo T. Farnsworth had become interested in television as a teen-ager, reading avidly about the most recent developments in photoelectricity and the development of the cathode-ray tube. In 1925, at the age of nineteen, Farnsworth, then a student at Brigham Young University in Utah, interested a group of businessmen from California in the commercial possibilities of television.

He moved to San Francisco and began to develop a system of his own. By 1930, with progress slow at best, Farnsworth was talked into moving to Philadelphia, where the Philco Corporation subsidized him for two years. It became evident that development would be painstakingly slow, and by 1932 financing had reverted to Farnsworth's California contacts only.

By 1935, Farnsworth had developed a system that was roughly comparable to that of Zworykin. Continued efforts to improve his system were not significantly successful. The basic camera tube—developed by Farnsworth in his teens—is called an *image dissector*. It is less sensitive than the iconoscope, which stores charge as long as an image is being projected on the face of the tube. The dissector, on the other hand, amplifies only for an instant during each scanning cycle. It produces an image of extremely high quality, however, and found some use in special industrial and military systems.

In developing his television system, Farnsworth took out over fifty patents, some of which are basic to modern television, including methods of synchronizing the transmission and reception and of *blanking*, which will be discussed shortly.

PRINCIPLES OF TELEVISION

Television works because it is possible to dissect an image into bits of information and to transmit them in sequence. To transmit a clear, sharply focused picture, several millions of such pieces of information must be sent in a second. This can be done because the electron beam in a transmitting or receiving picture tube is extremely narrow and can be controlled precisely by varying the electric and magnetic fields through which it passes. The action of these electrons on photosensitive or photoconductive materials produces a series of pulses of electrical energy whose amplitude is proportional to the light that has fallen on the surface being illuminated by the incident electrons.

The motion of the electron beam across an image is called *scanning*. It is controlled electronically. The electron gun, which emits electrons from a heated cathode, is stationary. The beam of electrons it produces sweeps the image in the way a page of text is read by the human eye, sweeping from left to right and dropping down after each line. The electron beam is swept electronically, by varying the magnetic fields surrounding it with magnetic coils outside the tube.

A complete sweep of the image is called a frame. The number of frames generally depends on the electric power frequency used in an area. For example, in the European area, where 50-Hz power is supplied, the number of frames transmitted per second is 25. In the United States, Japan, and Canada, 60-Hz power is used, and 30 frames per second are transmitted. Actually, the transmission of 25 or 30 complete frames per second would result in objectionable flicker; the eye is highly sensitive to changes in an image that occur too slowly. To avoid flicker the screen must be illuminated a minimum of 50 times per second.

Flicker is eliminated without increasing the number of complete frames per second by using a technique called *interlace scanning*. The scanning beam skips alternate lines as it sweeps downward across an image. The first reading, therefore, illuminates lines 1, 3, 5, 7 . . . , the second reading lines 2, 4, 6, etc. Thus, each frame consists of two separate transmissions, one of odd, the other of even lines. Each transmission is called a *field*. Two fields are transmitted per frame. In the United States, therefore, the field frequency is 60, which eliminates flicker entirely, and the frame frequency is 30, which is fast enough to give the illusion of continuous motion.

The total number of lines transmitted per frame depends on the system being used. Four different systems are currently in use. These are a 405-line, 25-frame system in the United Kingdom; a 525-line, 30-frame system in North and South America and Japan; a 625-line 25-frame system in Europe, Africa, and parts of Asia; and an 819-line 25-frame system in France.

The scanning spot moves from left to right along each line. When it reaches the end of the line it returns to the left side of the screen, at a point slightly below the previous line. On its journey back to the left of the image, it would produce bright lines across the transmitted picture. During this phase of the transmission, therefore, the scanner spot is "blanked"; the electron beam is "turned off" electronically. Similarly, when the spot has reached the lowest line of the picture it is returned to the left side of line 1, at the top of the picture, to start another frame. In this interval, another blank period occurs. During blanking periods, *synchronizing* pulses are transmitted. These supplementary signals are sent out by the transmitter. They are not part of the image being transmitted, but are an integral part of a successful transmission. They synchronize the scanning spot at the transmitter and receiver so they move at precisely the same instant. Otherwise the picture would be distorted. The average viewer knows the condition called "out of sync." At the end of each field a *vertical synchronization signal* starts the scanners at transmitter and receiver operating back on line 1 again.

Several improvements in television camera tubes have been made since the invention of the iconoscope. The first of these, the *orthicon,* was similar to the iconoscope but did away with the awkward positioning of the electron gun at an angle by using an extremely thin metal signal plate through which the image is focused onto the mosaic. At the opposite end of the tube an electron gun scans the mosaic. The beam is deflected horizontally and vertically by focusing and deflecting coils positioned around the tube.

The *image orthicon* is an extremely sensitive device that makes live, on-the-spot television transmission from remote locations possible. It is similar to the orthicon tube, but it also contains an *electron multiplier.* The photosensitive target screen arrangement is more sophisticated than in the iconoscope or orthicon.

The electrons from the gun strike the target with relatively low energy and are reflected back toward the gun in proportion to the charge on the target. The return beam of electrons depends on the charge on the target at each point of the scanning pattern. It therefore constitutes the signal current. This current passes through a five-stage electron multiplier. Electrons are emitted liberally in each stage, resulting in a greatly amplified current—a stronger picture signal. The final stage of the multiplier is fed into a conventional amplifier, which increases the strength of the signal still further.

One other tube uses *photoconductivity*—changes in electrical resistance in a material under the influence of light. In the *vidicon* the tar-

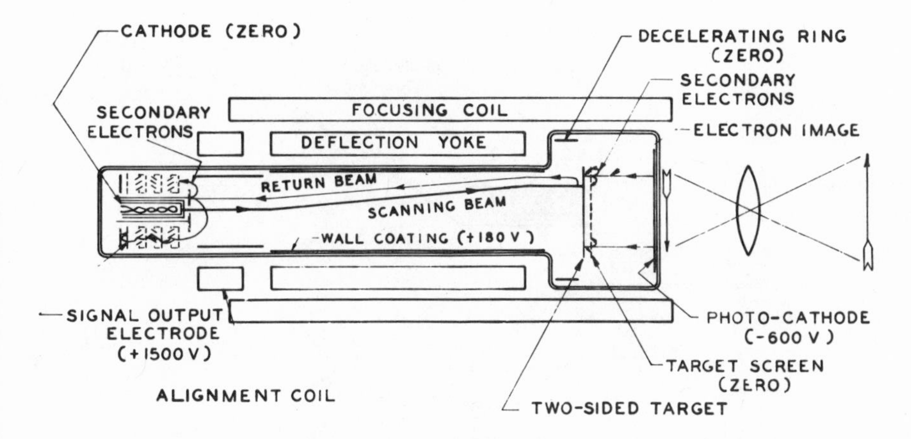

Fig. 59. Cross-section detail of the image orthicon.

get screen is composed of two layers of material: one layer is a transparent metal plate. Bonded to this plate is a photoelectric compound, composed principally of silicon. The electrical conductivity of this compound varies, depending on the amount of light that falls on it.

The optical image is focused at the end of the tube opposite the electron gun. It passes through the metal plate onto the photoconductive surface bonded to it. At each point on this surface the electrical resistance of the plate varies with the intensity of the light falling on it.

The electron gun bombards the plate with electrons, which scan the plate in the conventional manner. The charge flowing in the target plate depends on the resistance of the photoconductive coating on the metal plate. This depends on the intensity of the light falling on it from the image. The metal target plate is connected to an amplifier, which increases the strength of the signal output of the tube.

At the receiver, the picture information from the transmitter is fed to the electron gun in the kinescope (picture tube) (Figure 60) where the current in the electron beam depends on the strength of the signal being received. The beam is directed toward the screen, coated with *phosphors* which emit light in proportion to the intensity of electron bombardment. The beam, scanning these phosphors in synchronism with that of the camera tube, builds up, element by element, a reproduction of the transmitted image. Other signals are sent to other parts of the receiving circuit to keep the transmitted and received images synchronized.

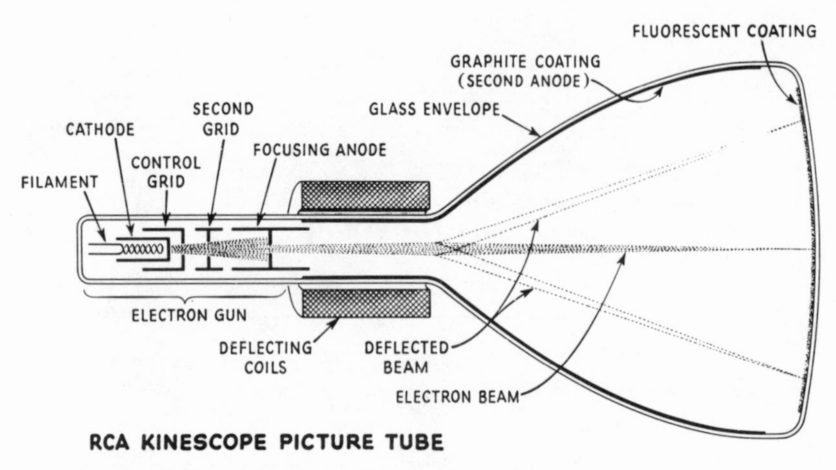

Fig. 60. A kinescope, or picture tube. *Courtesy of RCA.*

The speed at which picture transmission takes place is extremely rapid. For example, a 525-line picture, repeated 30 times per second, means that over 15,000 lines of picture information are transmitted each second.

Only a minute portion of the receiving tube screen is illuminated at any given instant. Yet the entire screen appears to have a picture on it because the phosphors lining the kinescope screen glow briefly even after the scanning beam has passed, and the eye retains images for about a tenth of a second after they have been removed. (Attempts to photograph a TV picture with exposures less than 1/30 second produce some interesting effects!)

While picture information, including the video, blanking, and synchronization, are handled by one part of the television channel, still another part of the channel is handling sound information. Sound is transmitted with entirely different equipment—microphones, transformers, amplifying, oscillating, and mixing circuits. It is only after the two signals have left their respective transmitters that they are combined (and "diplexed") and transmitted by the same antenna.

The standard television channel, video and audio, is 6 MHz wide, compared to a width of only 10kHz for the standard radio broadcast channel. The TV channel is therefore 600 times as wide as a broadcast channel. So much spectrum space is needed because a great deal of information must be transmitted in a very brief period of time, and this cannot be done in a narrower bandwidth.

Because of the wide bandwidth required by TV, it was originally assigned space in the VHF portion of the spectrum (54–216 MHz). But because of crowding, only a total of twelve channels could be allocated without seriously disrupting other services already assigned space in the 30–300 MHz portion of the spectrum.

To solve this problem the FCC allocated 70 additional channels in the UHF portion of the spectrum (300–3000 MHz). A total of 82 channels have been allocated in the United States:

CHANNEL NO.	FREQUENCY RANGE
2–4	54–72 MHz
5–6	76–88 MHz
7–13	174–216 MHz
14–83	470–890 MHz

Since propagation characteristics in the VHF and UHF portions of the spectrum are different, the UHF channels are not as desirable as those in the VHF band. For the same amount of power, the signals are weaker and subject to greater interference. Obstructions such as trees and buildings often result in signal losses in the UHF bands. These are not observed, or are not as great, in the lower portions of the TV spectrum. Because of this, the UHF channels are not only more difficult to operate, but frequently present a greater financial burden as well. For example, where an extended area is to be covered, low-cost repeaters, which pick up the TV signal and retransmit it, are used. Because the normal coverage area of UHF is less than VHF, and is more subject to interference effects caused by obstructions, more repeaters are required for a given coverage area for UHF than VHF.

COLOR TELEVISION

The first color television systems were mechanical and consisted of revolving color filters driven by a motor. A mechanical system, developed and supported by CBS, was approved in 1950 by the FCC. One of its primary drawbacks was that it was incompatible with black-and-white TV—color transmissions on the CBS system could not be received on a conventional black-and-white receiver.

RCA had been working toward the development of a compatible color system, and the FCC decision to adopt the CBS system led to the acceleration of the RCA program to develop compatible electronic color TV. David Sarnoff, who said, on hearing of the FCC ruling, "We may have lost the battle, but we'll win the war," put his engineers on an eighteen-hour day and a virtually unlimited budget. Within three years, the history-making three-gun color tube had been perfected, and the FCC, reversing its original ruling, adopted the compatible electronic color system whose development Sarnoff had so diligently pursued.

The electronic color television developed by RCA engineers actually consists of two separate transmissions, the luminance, or brightness, and the chrominance, or color information. (When received on a black-and-white receiver, the chrominance component has no effect.) This remarkable achievement made it possible to proceed to the next step in television development without making the millions of black-and-white receivers already on the market obsolescent.

The color television camera developed by RCA engineers contains three image orthicon tubes. The image to be televised is passed through a lens system, which breaks it into three identical images with the aid of four specialized mirrors. These are fed into the three tubes. The "dichroic" mirrors through which the images pass are color-selective, passing light of one color and reflecting all other light. One set allows only blue light to pass through to its image orthicon. The second set passes only red light, the third, green. Through the proper combination of these primary colors virtually any color can be duplicated.

When the image has passed through the mirrors, three corresponding sets of signals, one for each of the primary colors, are fed out of the image orthicons. The signals are then combined into two separate signals, one carrying the brightness information, the second the color. The brightness information is formed in a circuit that simply adds the signals from the three orthicons. The signal carrying the brightness information can be received on a black-and-white receiver, which will respond as if no color information were present. This makes the system compatible.

Part of the blue, green, and red signals are modified and fed into a second circuit, which transmits the chrominance information as part of a second set of signals. In a black-and-white receiver, these signals have no effect. In a color receiver, they are picked up by specially designed components that feed them to circuits which break them into red, blue, and green signals, exactly as they were transmitted. They are then fed to the color kinescope, which contains three electron guns, one for each of the primary colors.

The secret of color reception lies in the remarkably precise screen of the receiving tube (Figure 61). The screen (in the conventional shadow-mask tube) is coated with phosphor dots, arranged in groups of three, one green, one blue, and one red. If the screen is viewed through a microscope when the set is on, each of these dots can be seen easily. The inside of the screen is covered by a mask, which contains holes, one in front of each set of three phosphor dots. Electrons from the three guns in the receiving tube are fired through each hole in sequence. The angles at which the three guns fire are slightly different, so that the red gun beams electrons only toward the red phosphor dot behind each hole. The blue gun fires only at the blue phosphor, and similarly the green. Thus, the image that was broken up and

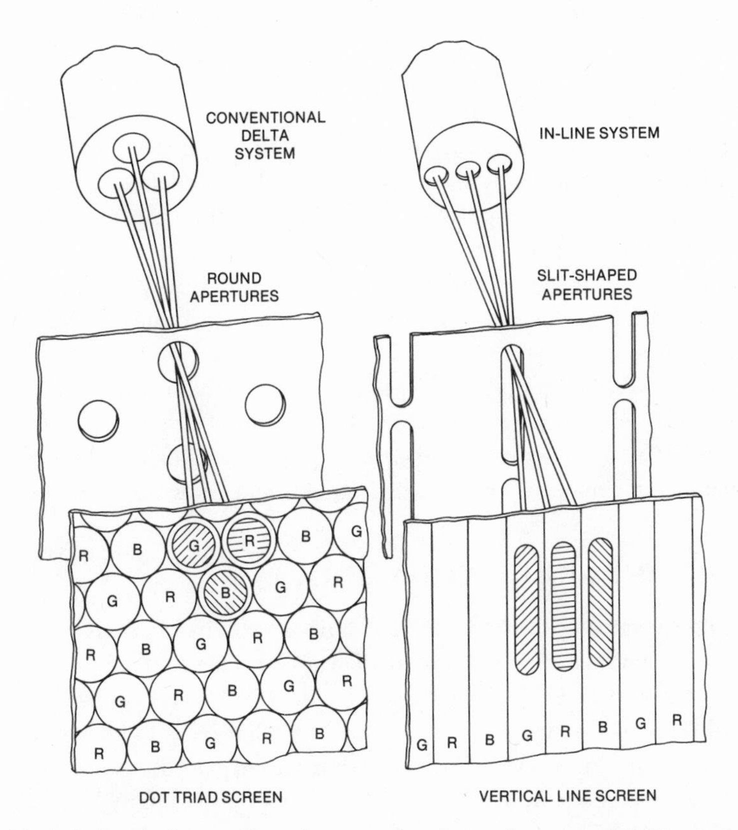

Fig. 61, A & B. Shadow-mask and screen detail: A—the original RCA tricolor dot mask and screen with discs in delta formation; B—the in-line system.

fed to three separate image orthicons in the transmitting tube is now reconstructed by the proper combination of the original primary colors.

The dots illuminated by the separate electron guns—although they show up well with a magnifying glass—are too small to be distinguished by the human eye, which sees only the net result of the complex scanning taking place at the rate of more than 15,000 lines per second. The eye of the viewer actually mixes the colors illuminated in each set of dots.

The shadow-mask tube is a marvel of twentieth-century mass production methods. A 21-inch tube, for example, contains over one million precisely placed phosphor dots and some 350,000 holes. Dots and holes must be arranged so that the stream of electrons from each of the guns in the receiving kinescope impinges only on the proper dot.

The extreme complexity of the three-dot color tube sometimes results in relatively poor picture quality. To fire properly, the three beams from the guns—in the older and probably still most common type of tube—require some twelve different covergence circuits. These are not only complicated but are difficult to adjust. As a result, the 1970s have seen the introduction of two new receiving tubes. RCA still uses three guns, but instead of masked groups of three phosphors, the screen is composed of vertical strips of color phosphors. These are very thin and make it necessary to align the electron beams in one direction only, the horizontal. With this configuration, much simpler convergence circuitry is used. The electron guns, which in the earlier tubes had been arranged in the form of a triangle, are placed in a straight line.

A second tube, introduced by the Sony Corporation, is the *Trinitron*. It uses a single gun to illuminate successively vertical strips of phosphors. In both the RCA and Sony systems, the complex convergence circuitry is eliminated, making adjustments less frequent and color more stable.

FREQUENCY MODULATION (FM)

The carrier wave fed to the transmitting antenna of a radio or TV station can be modulated in a number of ways. In standard broadcast transmission, the carrier is amplitude modulated (AM) by varying the strength, or amplitude, of the signal. Thus, if a tone of 1,000 Hz is to be transmitted, the stength, or amplitude, of the carrier is varied one thousand times per second. How much it would vary would depend on the loudness or softness of the 1,000-Hz note.

Edwin H. Armstrong, pioneer and innovator in the radio field, had long grappled with the problem of overcoming static and interference in radio broadcasting. He worked on the problem from the early 1920s and by 1933 had developed a system of *frequency modulation* (FM) which eliminated or greatly reduced many of the forms of interference that had plagued standard (AM) radio transmissions.

If a 1,000-Hz note is transmitted by FM, the amplitude of the signal remains constant throughout the transmission, but its *frequency* is shifted up and down 1,000 times a second from the nominal frequency of the carrier wave. FM is graphically represented in Figure 62. The *amount* of swing depends on the strength of the signal—in American broadcast FM transmissions the maximum signal can swing 75 kHz each way from the center frequency. The *bandwidth* is said to be 150 kHz.

Armstrong went to RCA with his FM system, and tests were begun in 1934. Although generally successful, there were some problems with fading and interference, and Armstrong pushed for continued tests with higher power. RCA, heavily committed to developing a commercial television system, balked at further significant expenditures in developing FM. There were many reasons for this, not the least of which was the fact that television and FM would compete for frequencies in the same region of the spectrum, both being most effective in the VHF range, and that this competition would ultimately extend to listenership.

Armstrong, whose innovations in radio had made him a wealthy man, decided to go it alone; conducting research using personal funds, he applied to the FCC for a construction permit. The FCC in 1944 allocated 40

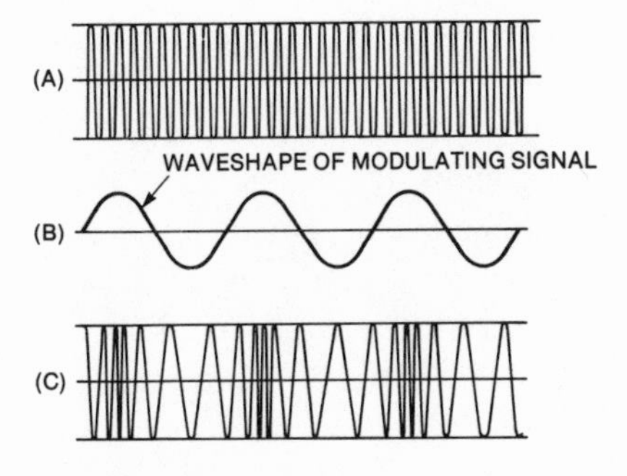

Fig. 62. Frequency modulation illustrated. A—the carrier wave; B—the modulating signal; C—the frequency-modulated wave. Its amplitude remains unchanged, whereas its frequency varies with the intensity of the modulating signal. *From* The Radio Amateur's Handbook.

channels in the 42–50 MHz region of the spectrum. These consisted of 35 commercial, and 5 noncommercial educational channels.

On October 31, 1940, the commission granted construction permits for the first fifteen FM stations. By the end of that year there were ten more. Then all radio construction was frozen during World War II, though more than forty prewar FM stations continued to serve some 400,000 receivers.

During the war, a controversy over assignment of frequencies to FM and TV arose, with both interests vying for the lower portion of the VHF bands. Hearings were held in 1944, and on the basis of testimony by a technical expert, who asserted that sky-wave interference on the lower frequencies would degrade the quality of FM service, the commission reallocated the bands previously assigned to FM, moving the service "upstairs" to the frequency range of 88–108 MHz. The number of channels was increased to 100, including 20 in the range 88–92 MHz for noncommercial educational use. In spite of the fact that more channels were made available to FM, it was dealt a crippling blow, because it made all the equipment, transmitters, receivers, antennas, etc., that had been built and sold for the old FM bands obsolete.

Armstrong, recognizing the seriousness of the setback, continued to fight, and in 1947 succeeded in getting the technical expert who had testified about the propagational shortcomings of FM in the lower bands to admit that he had been in error. In spite of his efforts, however, the FCC assignment stood. But although FM even in its new surroundings attracted a loyal following of listeners—particularly among music lovers—it did not immediately become as popular as AM broadcasting. Many FM stations failed, and the number dwindled steadily until the mid-1950s.

His funds depleted, his energy drained, his will to fight gone, Armstrong, after years of struggle to return FM to its original frequency range, committed suicide in 1954.

It is ironic that within a few years after his death FM enjoyed a great upsurge of popularity. This was caused by several factors: in 1955 the FCC authorized FM broadcasters to transmit supplementary "store-casting"—programs intended for stores, doctors' and dentists' offices, banks, etc. These programs usually contained no commercials and were purchased by the stores for a fixed fee, which included installation of the receiving equipment. "Store-casting" provided a supplementary means of income for FM broadcasters. (The programs are carried on a subcarrier, which is sent out on the same channel, but on a frequency too high for the human ear to detect. This subcarrier can be picked up and decoded by specially constructed receivers.)

The second factor in the rebirth of FM popularity was the increased interest in high-fidelity music. The FM channel is 200 kHz wide (150 kHz for program, and 25-kHz "guard bands"), compared to a width of 10 kHz for AM. The wider range of the FM channel permitted the transmission of programs whose audio range was up to 15,000 Hz, which comprises virtually the entire range of human hearing, as opposed to a much narrower range (nominally 5,000 Hz, but often exceeded) on AM. The third, and probably most significant factor in the recent success of FM is stereophonic boadcasting. In 1961 the FCC approved for FM broadcasting the transmission of multiplexed signals—the transmission on a single channel of the two signals necessary to reproduce music stereophonically. This assured the success of FM.

16

Radar and Semiconductors: Stepping Stones Toward Space

It is unfortunate that many of man's outstanding scientific achievements have been made while developing instruments of destruction or defense. Two of the most remarkable scientific achievements of the twentieth century—the development of atomic energy, and the exploration of space—illustrate that fact.

Without rocket fuels man could never have gone very far above the surface of his planet. But even with those propellants, the successful exploration of outer space would have been difficult indeed without the radio tools to track, to guide, to obtain data, and to communicate with space vehicles.

One of the most important of these radio tools, radar, was developed in 1935. Robert Watson-Watt of Great Britain's National Physical Laboratory was asked whether it was feasible to develop a beam of damaging radiation that could be used against enemy aircraft. Not being too much impressed with the idea of working on radio death rays, Watson-Watt proposed that efforts be directed instead to developing a method of detecting aircraft by reflected radio waves.

Within six months, aircraft to distances of 43 miles had been detected. The success of the initial experiments led to the recommendation that a chain of radio-detection stations ringing the British Isles be constructed. Funds for the project were allocated, and the first operational *radar* network was established.

The principles of radar are simple and had been known for years: radio waves are reflected by objects such as mountains, ships, and aircraft.

At higher frequencies (VHF, UHF, and SHF), even clouds or birds reflect radio waves. The waves are sent out by a powerful transmitter, strike an object, and are scattered in all directions. Some of the energy is reflected back toward the transmitter and can be picked up on a sensitive receiver and displayed on an oscilloscope. Knowing the speed at which the waves travel, the time between the transmission of the radio wave and its return to the receiver indicates precisely the distance of the object from which the radio wave was reflected. The return signal can be displayed as a luminous dot, and if the transmitting and receiving antennas are directional and rotate, and the equipment is adjusted so the center of the oscilloscope represents the center of the transmitting antenna, the direction of the object can also be determined.

Scientists and engineers working with radio waves frequently observed radio echoes, or reflection phenomena. In the June 1900 issue of *Century* Magazine, Nikola Tesla proposed that a moving object such as a ship could be detected, and its motion plotted, by reflected radio waves. In 1904 a German engineer, Hulsmeyer, patented a simple device that used radio echoes to prevent the collision of ships at sea. There was little interest in Hulsmeyer's work at the time, and nothing came of it. Hugo Gernsback, in his science-fiction novel *Ralph 124C41+*, published in 1911, described and illustrated an apparatus (used to locate a fleeing spaceship) that paralleled modern radar so closely that the description and diagram can be used to explain radar to a student. But there was no use for such a device at the time, and even the author forgot about it until it was called to his attention by a newspaper reporter during World War II. In 1922, Marconi proposed a similar system during a lecture to the Institute of Radio Engineers in New York City.

Reflected radio energy was first used to explore the ionosphere. Scientists Breit and Tuve, working in the United States, sent very short pulses of radio energy in a vertical direction, and by timing the return pulses were able to demonstrate the existence and distance of reflecing layers in the earth's upper atmosphere. Appleton and his associates, working in Great Britain, used continuous waves of varying frequency to arrive, almost simultaneously, at the same conclusions as had Breit and Tuve.

In the 1930s scientists in Germany, France, and the United States engaged in ship and aircraft detection methods and altitude indicators, using reflected radio waves, but the first integrated system was developed and organized under the leadership of Watson-Watt in England.

During World War II the development of radar was accelerated, with the United States and Great Britain working hand in hand, sharing information, mutually developing more sophisticated methods of detecting and tracking aircraft, ships, and submarines. The term *radar* was coined in 1942 by U.S. navy scientists, although it remained "top secret" and did not reach the public until 1943. It is an acronym for radio detection and ranging. Part of the cooperative operation between the two countries was the introduction, in the United States, of the cavity magnetron, a device developed in England which revolutionized the generation of microwave radio frequency energy.

TYPES OF RADAR

The most common type is pulse radar. Short bursts of radio frequency energy are fed to an antenna, which beams the signal outward. If the beam strikes a target within the radiation pattern of the antenna, part of the signal is reflected and detected by a receiver close to the transmitter. The signal is amplified and displayed on a cathode-ray tube, like a television signal.

The burst of energy the transmitter generates is masked at the receiver by a TR (transmit-receive) box, which protects the receiver from being destroyed by the powerful emission sent out by the transmitter. The TR box, in effect, shuts off the receiver while the transmitter is in operation. Once the pulse has been sent, the TR switch shuts off, activating the receiver.

The system, then, consists of alternately sending a pulse and receiving its echo. The cycle is then repeated and another pulse is sent, during which time the receiver is shut off, after which the transmitter is inactive and the receiver listens for the returning signal. The number of pulses per second varies from just a few to several thousand, and is known as the *pulse repetition frequency* (prf). The prf depends on the type of object being tracked and its distance from the transmitter. The period between pulses must be longer than the time it takes to reach the object being tracked and the return trip to the receiver. Otherwise, the next transmitted pulse would mask the return impulse at the receiver. The pulse width is determined by the minimum distance of the object being tracked. For example, a radio wave travels approximately 1,000 feet (300 meters) in 1 microsecond. In a 1-microsecond pulse-width radar system, the object being tracked must be at least 500 feet (about 150 meters) from the transmitter. Otherwise the pulse will still be on when the wave has returned to the receiver. Most navigational radars, therefore, operate with a pulse width of 0.1 microsecond, so that objects very close to the transmitter can be tracked.

Since it is desirable for all the echoes from a single pulse to be returned before the next pulse is transmitted (otherwise the measurement of the range will be ambiguous, because it will not be clear which pulse caused which echo), the prf is determined by the maximum distance of the object to be tracked.

For example, if a pulse is 1 microsecond wide, and the prf is 250, then the period between pulses is 4,000 microseconds. In that length of time a radio wave travels some 656 miles, making the maximum range of 250-pulse-per-second radar a little less than 328 miles. The operator of a radar set can adjust both the pulse width and the prf to change the ranges over which the equipment will operate.

Pulse radars employ highly directional "dish" antenna systems. Both the transmitting and receiving antennas used in radar systems have extremely narrow beam widths, and can pinpoint objects with great preci-

Fig. 63. The 150-foot-diameter parabolic antenna that tracks missiles on the Kwajalein Missile Range. *Courtesy of General Telephone and Electronics Corp.*

sion. These antennas usually rotate, and are able not only to determine the range and direction of an object, but its elevation (in the case of planes) as well. Elevation is obtained by varying the vertical angle to that at which the antenna signal is maximum.

DOPPLER RADAR

A type of radar that is of great value in tracking moving objects is the Doppler or CW radar. Some of us have observed that an automobile horn or train whistle sounds higher pitched when it is moving toward us, lowering as it passes and moves away. This principle, first described by an Austrian physicist, Christian Doppler (1803–1853) is due to changes in frequency caused by the motion of the source of the sound toward or away from the hearer. The Doppler shift also applies to electromagnetic radiation, and radio signals reflected from an object moving toward the radar transmitter are higher in frequency than signals reflected from one moving away from it.

In Doppler radar, part of the transmitted energy is fed to the receiver, so that when the reflected signal arrives at the receiving antenna it can be compared with the original signal. The difference between the original signal and the reflected signal is displayed on a radarscope, and the operator can read out directly whether the object being tracked is moving toward or away from the observer, and at what speed.

The range (distance) to an object being tracked can be determined by sending frequency-modulated CW signals. The frequency being sent by the transmitter is varied at a uniform rate and part of it is sent to the receiver, where it is compared with the signal being reflected. Since it takes time for a signal to travel from the transmitter to the reflection point and back, the signal arriving at the receiver will be slightly different from the one being fed directly to the receiver. Both signals are combined in a mixer, which produces a beat frequency that depends on the distance to the object being tracked. The distance to a rapidly moving object cannot be determined by FM CW radars because the beat frequency changes too rapidly, due to Doppler shifting.

APPLICATIONS

The most important military application of radar is that of scanning for aircraft, ships, and missiles. Modern defense systems depend on early warning of approaching hostile craft, and radars, using electromagnetic radiation that travels at the speed of light, are vital in such detection systems. Radars are also used to control the firing of weapons. Radars aboard aircraft, for example, can track another aircraft and supply position and speed information to a computer, which calculates its path and aims and fires a weapon. Radars are also used to fire antiaircraft guns after tracking enemy aircraft.

Nonmilitary applications of radar have been growing in use and importance. All commercial aircraft and ships are equipped with radar, which helps locate other nearby objects, such as ships or planes, icebergs, or land. Radar on planes assists in determining altitude above the ground and the presence of other aircraft and mountains in the vicinity; radar at airports assists in landings during poor weather conditions and during crowded hours when other aircraft are in the area.

Reduced costs—largely due to integrated circuitry—will probably extend civilian applications of radar to small craft, security systems, and eventually automobiles.

Meteorological radars have significantly increased forecast accuracy, especially for the short term. Storm centers and squall lines are routinely tracked by Weather Bureau radars, enabling the rapid dissemination of storm forecast information. These services have proved especially useful in predicting tornadoes. They are also supplied to airlines to warn pilots of trouble zones to avoid.

RADAR ASTRONOMY

Radar studies of the solar system have provided valuable astronomical information about the sun, moon, and nearby planets. Radar astronomy began in the 1930s with the observation that signals could be reflected from the ionized trails left by meteors when they entered the Earth's atmosphere. These studies provided useful information about the behavior of meteors in the vicinity of the Earth.

The first signals were bounced off the moon in 1946. Subsequent studies of the behavior of the reflected radio waves provided valuable data about the moon's surface. In 1958, scientists at MIT's Lincoln Laboratory succeeded in reflecting radio signals from the planet Venus. Since then, the sun and all the planets to Saturn have been probed with reflected radio waves. These studies have been of great value to astronomers.

One of the major applications and uses of radar has been in space science, where radars are used in launching and tracking Earth satellites, space probes, and manned satellites. Exploration of the moon and the Skylab project would not have been possible without radar, which gives extremely precise information about the speed, direction of motion, and orbital characteristics of space vehicles and probes.

MICROWAVE TECHNOLOGY

Early radars employed frequencies in the very-high-frequency portion of the electromagnetic spectrum (frequencies between 30 and 300 MHz). As radar became more useful and important, ultra-high frequencies were used. The reason is simple: the size of the antenna in a communication system is proportional to the wavelength of the signal. At 30 MHz, for example, the wavelength is 10 meters, and an antenna with half-wave elements must be about 16 feet long. To obtain directivity, either parabolic dish-type reflectors or multiple-element antennas are required. To install a VHF system on an airplane or to supply it to combat forces in the field would make for cumbersome logistics.

At 300 MHz, on the other hand, the wavelength is 1 meter, and at 3,000 MHz, it is 10 centimeters, or about 4 inches. At these dimensions, portable radars become feasible. As early as the late 1930s intensive efforts were being made to generate frequencies in the microwave portion of the spectrum. The term *microwave* applies to frequencies from 1,000 MHz (1 GHz) to 300,000 MHz (300 GHz), which is the beginning of the infrared region of the electromagnetic spectrum.

But conventional vacuum tubes were inefficient or inoperable at microwave frequencies, and efforts were made to develop special tubes that could generate high power at microwave frequencies. The problem was solved in Great Britain during the late 1930s with the invention of a new type of tube, the *cavity magnetron*, in Figure 64. One of the most important elements in relatively low-frequency circuits is the tank, or resonant, circuit. It consists of one or more inductances and capacitances that together form a closed, resonant circuit; for example, the coil and capacitor in the plate circuit of a tube transmitter. The cavity magnetron is based on the fact that certain types of cavities have their own characteristic electromagnetic resonance. A physical analogy can be drawn with an organ pipe. Depending on its dimensions, each pipe has a characteristic acoustical resonance. In the cavity magnetron, each cavity resonator has its own

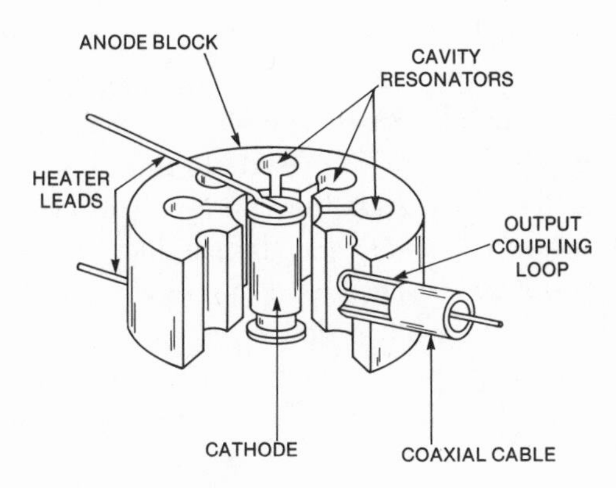

Fig. 64. Details of the cavity magnetron.

particular resonant microwave frequency, depending on its dimensions.

The cavity magnetron consists of a series of cavities in a direction parallel to their cylindrical axes. Electrons from the cathode flow toward the anode, but the magnetic field makes them move in circular paths en route to it. Depending on the dimensions of the cavity, the anode voltage, and the strength of the magnetic field, the electrons can be made to travel so that they just graze the openings at each cavity. This motion causes the cavities to oscillate at their natural microwave frequency, and the microwave power thus generated is fed through an output slot in one of the cavities.

The cavity magnetron was the first practical high-power source of radio-frequency energy at frequencies above 1,000 MHz. It greatly extended the usefulness of radar as well as microwave communication systems in the lower microwave regions. The magnetron has two drawbacks. At the higher microwave frequencies the cavities have to be so small as to make the tube hard to manufacture, and the small tube cannot deliver much power. Because a powerful magnetic field is needed, a heavy electromagnet, whose weight increases with the power output of the tube, must be used. This makes the magnetron impractical in portable or mobile radars. In airplanes, for example, high-power magnetrons add considerable weight to the craft, reducing its utility.

The *klystron*, another microwave oscillator, needs no external magnetic field, making it practical for use in portable radars and in situations where light weight is an important consideration. The klystron operates on a different principle from that of the cavity magnetron, depending on a series of grids that speed up or slow down the electron stream from the cathode to the plate, and take power from the stream as they slow it down.

The klystron can sometimes be made to function as an amplifier by feeding energy to the cavity grids from an input resonator. The signal is amplified in an output resonator, and the amplified output signal travels into a waveguide through an aperture in the resonator.

Klystrons have been built to deliver as much as 10 million watts of pulsed power and 100,000 watts of continuous power. They can operate at frequencies up above 100,000 MHz, but at these frequencies the power output is low.

Another specialized microwave tube, which delivers much less power than either magnetrons or klystrons but which can amplify over a much wider range than either of the others, is the *traveling wave tube* (TWT), which consists of a coil of wire wound around a long, narrow, evacuated tube. An electron gun fires a beam of electrons down the center of the tube, to a collector at the other end. An input waveguide (see below) sends a signal into the helix, whose pitch is adjusted so that the signal and the electron beam travel at the same speed along the tube. In traveling through the helix, the signal generates an electric field, which interacts with the electrons, causing them to bunch. The bunched electrons, in turn, transfer their energy to the helix, as a result of which the signal is amplified as it travels along the tube. By the time it reaches the end of the tube it may be amplified as much as 100,000 times. It is then fed to a waveguide. Since no resonant cavity is involved, a wide range of frequencies can be amplified by a TWT. Depending on the signal fed into the helix, the TWT can be made either to amplify signals or to generate microwaves.

OTHER COMPONENTS

In addition to the special types of tubes needed to generate or amplify microwave energy, radio waves at frequencies above 1,000 MHz required that new technology be developed to transmit them from one point to another. For example, ordinary transmission lines do not conduct microwave energy efficiently. The effects of inductance and capacitance, even in wire, become significant because they vary with frequency. The greater the frequency, the more serious these effects can be, to the point that no energy is transmitted. It therefore became necessary to develop special cables—coaxial transmission lines and waveguides—to transmit the energy from the transmitter to the antenna, and from the receiving antenna to the receiver.

The coaxial transmission line consists of an inner conductor surrounded by an insulator, generally of some plastic material, and an outer cylindrical conductor. Coaxial transmission lines can efficiently carry frequencies up to several thousand MHz, at relatively low powers.

Power at frequencies above 500 MHz is commonly conducted from transmitter to antenna, and from antenna to receiver by a *waveguide*, a precisely machined tube, either circular or rectangular, within which the microwaves travel. The range of frequencies a waveguide propagates efficiently depends on its dimensions, and precision tooling is a must for efficient operation. By proper design, rectangular waveguides can be bent into complicated paths, including right angles, and the signal continues to travel effectively through them. One drawback to the rectangular waveguide is that at high frequencies the dimensions of the guides become so small as to make them difficult to handle and to manufacture. Even more important: the power-handling capacity of waveguides operating above 100,000 MHz diminishes drastically.

Because microwaves lie in the region of the electromagnetic spectrum between the longer radio waves and light, they have certain features common to both. At the upper end of the microwave region the lightlike characteristics of the radiation must be taken into account in designing antenna systems. Like light, microwaves can be beamed, reflected, and focused.

Because of the vast amount of spectrum space in the microwave region, their importance to communications has become increasingly great. The microwave portion of the spectrum (1–300 GHz) encompasses some 299,000 MHz. This compares to approximately 1.5 MHz for the entire standard (medium wave) American broadcast band, and less than 1,000 MHz for the entire electromagnetic spectrum below it. A single microwave channel can carry hundreds of telephone conversations, television pictures, and data simultaneously.

Their lightlike qualities make them useful for communication in space. The vast distances that have to be covered in communicating with space vehicles—both in the region of the Earth and in deep space—required extremely high-gain antennas. High gain is obtained by focusing electromagnetic energy into extremely narrow, accurately directioned beams. This can be done by designing lens systems, using either metallic or dielectric lenses, or by using horn and parabola-shaped reflectors to focus energy from a radiating element into an extremely narrow beam.

THE TRANSISTOR

The transistor has revolutionized electronics. It has become the single most important circuit element. Without it, men would probably not have walked on the moon; the exploration of the solar system, now under way through the use of satellites, would not have been begun; the mushrooming space-communication industry would have remained dormant. Without transistors the modern electronic computers upon which all space sciences depend would not have been feasible. It has made possible circuit functions that were too complex to be contemplated before its invention.

The transistor is a *solid-state* device, made up of semiconducting material, which performs in a solid most of the functions of a vacuum tube—it conducts, detects, modulates, amplifies, and controls electric current in electronic circuits. Moreover, it does so at lower cost and at a fraction of the power that a vacuum tube uses. It does not need a warmup period, resists shock and vibration, is extremely reliable, and when properly manufactured can run trouble-free for decades. Most important is its size. It is tiny, and can be made smaller than the head of a pin.

The development of transistor technology has led to the miniaturization of other circuit elements, and methods have been developed for depositing both capacitors and resistors side by side with transistors on chips of silicon. Fifty or more circuit elements can be deposited on a chip of silicon one millimeter square. These *integrated circuits* have further revolutionized the electronics and communications industries.

Although the transistor was invented in 1947, semiconductor crystals were used in the early days of radio both as rectifiers and detectors of radio signals. "Cat's whisker" detectors were used into the 1920s, until vacuum tubes became cheap enough and vacuum tube technology advanced enough to replace them. However, point-contact crystals continued to be used to detect UHF signals.

In the 1920s and 1930s experiments in which semiconductors were used to oscillate or to amplify currents were conducted in Russia by Lossev and in the United States by Dr. Julius Lilienfeld, who filed a patent in 1925 for "A Method and Apparatus for Controlling Electric Currents"; but a reliable amplifier was never built, because nothing was known about the importance of impurities in crystals.

During World War II, semiconductors had a rebirth in the communications field when crystal detectors were found to be more useful than vacuum tubes in detecting microwave radar signals. Still, at the end of

World War II no one could consistently make semiconductors amplify.

In 1945, Bell Laboratories made a crucial decision. The Bell system was exploding in complexity and size. The switch from manual to automatic exchanges prior to World War II had streamlined the operation and reduced labor costs, but the exceedingly complex electromechanical networks assembled in the automation process needed a large, costly staff of skilled maintenance crews. It was decided, therefore, to attempt to develop an entirely new switch—one that wouldn't wear out as quickly as the vacuum tube, or be as fragile or bulky or large or expensive.

Accordingly, the Bell system funded a program of basic research in search of a substitute for the electron tube. One team of Bell scientists became interested in semiconductors. They had had considerable experience with the intriguing crystals in their microwave radar researches, and the parallel use of crystals and vacuum tubes as detectors led some of the scientists to feel that the materials had possibilities as amplifiers of current.

Among the Bell researchers, William Shockley, John Bardeen, and Walter Brattain devoted their full time to research on semiconductors. One experiment led to another, and failure at first resulted in new experiments and new theories. Ultimately the experiments succeeded, and an entirely new phenomenon, the transistor effect, was discovered. Two pointed metal contacts, similar to the cat's whisker, were brought into contact with a germanium semiconductor material. When a small positive voltage was applied to one of the contacts, the flow of current in the other contact increased significantly. An amplification of 40 was achieved. The resistance of the second point contact was found, in this momentous experiment, to depend on the current flowing at the first one. The observation that the resistance was, in effect, transferable, gave rise to the name *transistor*, a *transfer resistor*. The three Bell scientists continued their research, and in 1949 Shockley proposed a new type—the junction transistor—and predicted what its major properties would be. Shockley's proposals were based on the discovery that the transistor effect took place in the body of the crystal, rather than along its surface, as had been originally supposed. For their work on transistors, Shockley, Brattain, and Bardeen received the Nobel Prize for Physics in 1956.

WHAT A SEMICONDUCTOR IS

As the name implies, semiconductors have electrical properties intermediate between those of conductors, such as copper and silver, and insulators, such as rubber or glass. In a conductor, electrons flow, or move from one atom to another, producing a current. In an insulator, the electrons are bound tightly to the atom, and even a high voltage does not result in current flow.

In a semiconductor the electrons are bound relatively tightly to the atom, but its grip can be loosened in a number of ways. One of these ways, discovered by Bell's Nobel winners, was by "doping" the material—adding impurities that altered its electrical characteristics. They further discovered that different types of impurities resulted in different types of conduction.

The reason that doping alters the characteristics of semiconductor materials lies in their atomic structure. Each atom has a central nucleus with a positive charge, and numbers of negative electrons around it. These electrons surround the nucleus in a series of *shells*, at different distances from it. We tend to think of the atom as a solid nucleus with electrons orbiting around it like satellites around the Earth. Each of these orbits can contain only a certain number of electrons. The innermost has a capacity of 2, the next two of 8 each, and the following two of 18 each. If the outside shell of each atom is completely filled, the atom is highly stable and does not enter into chemical combinations with other elements.

If the outer shell has a very small portion of its ideal number of electrons, or is just one or two electrons short of having a complete outer shell, it enters into combination very readily. Thus neon (atomic number 10), which has a complete outer shell of 8 electrons, shows no desire to combine with other elements, while fluorine (atomic number 9), one electron short of a complete outer shell, and sodium (atomic number 11), which has a single electron in a new outer shell, are among the most active materials known. The single outer atom of sodium often finds a home in the outer shell of chlorine (atomic number 17), which lacks one electron of a complete outer shell, thus forming the common sodium chloride, or table salt. (The atomic number of an element is the number of orbiting electrons around its center.)

The materials most used in transistors are silicon (atomic number 14) and germanium (atomic number 32). These have outer shells with 4 electrons. In a crystal, these atoms *share* their electrons. Thus a germanium atom in a crystal may share 4 electrons with 4 neighboring atoms, making what is called a cubical crystal lattice, a very stable material, through which any motion of electrons is difficult.

To dope a crystal, a very small amount of some material having one less or one more electron in its outer shell than the doped crystal is added to it. For example, an atom of arsenic is substituted for one of the germanium atoms in the crystal. Arsenic (atomic number 33) has five electrons in its outer shell, leaving one to circulate when the other four are bonded to neighboring atoms. Germanium that has been treated with arsenic, therefore, has an excess of electrons, or negative charges, and is called n-type germanium. These extra electrons are the basis of current flow in an n-type semiconductor. (Note that, although the crystal has an excess of electrons, it is still electrically neutral because for each extra electron in the crystal there is an extra positive proton in the nucleus of the arsenic atom.)

The element that gives the crystal its free electrons is called a *donor* impurity, because the extra electron can be donated as a *mobile charge carrier* to produce a current flow in the material if a voltage is applied to it.

Just as a crystal can be doped with arsenic, which has five outer electrons, so can it be doped with gallium, which has only three. The absence of the electron with which neighboring germanium atoms would like to hook up is called a *hole* (one of the normally shared spaces is empty instead of containing an electron). The gallium added to the crystal is called an *acceptor impurity,* because it accepts a free electron easily. The crystal formed by doping germanium with gallium is called p-type germanium,

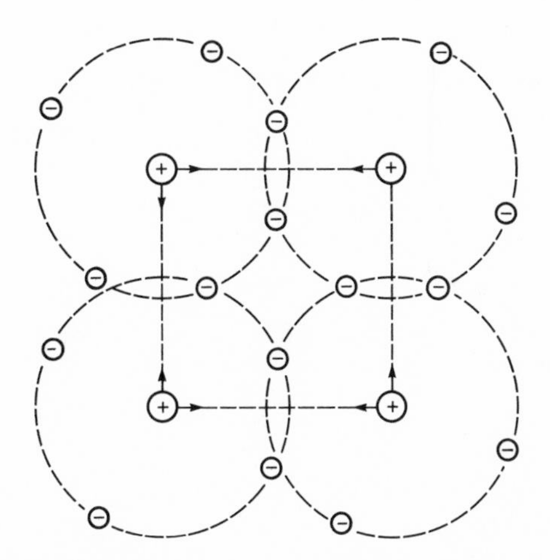

Fig. 65. Two-dimensional attempt to represent the cubical crystal lattice of germanium. Each of the four electrons in the outer shell (the "corner" electrons) is shared with the neighboring atom. (Inner shells are not shown.) The arrows between the center nuclei represent the bonding force that keeps the atoms together.

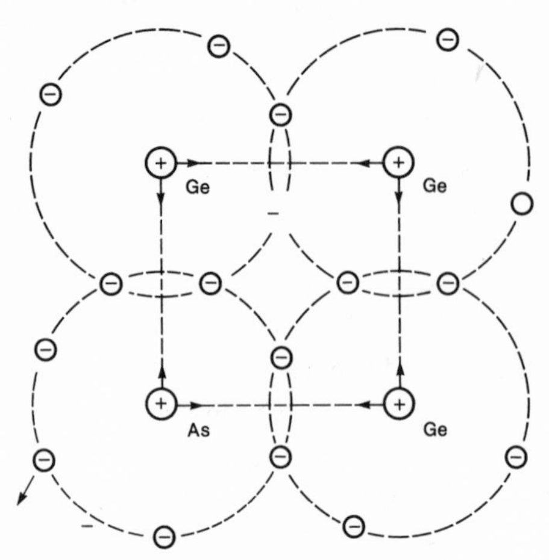

Fig. 66. What happens when an arsenic atom is inserted. An extra electron—which may want to get away—is added to the outer shell.

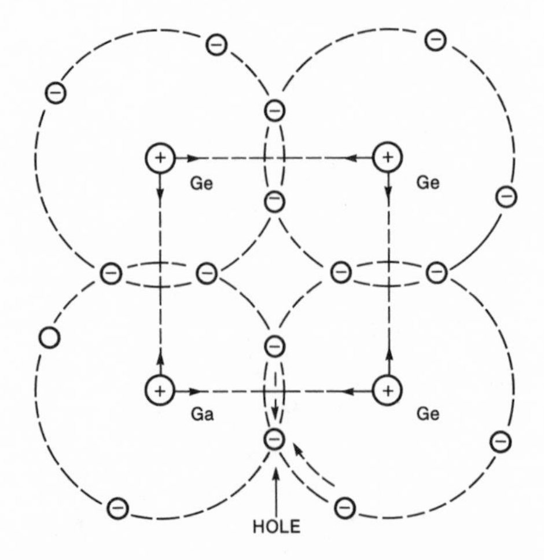

Fig. 67. The effect when the crystal is doped with gallium. One of the atoms has a "hole" in the outer shell, which an atom from a neighboring atom may want to fill. This can be a basis of current flow, though only a very small number of the atoms in the structure are of the "impurity" material.

because it has an excess of holes, which have the same properties as positively charged particles.

Extremely small quantities of impurity material are used in the doping process—less than one part per million. This requires extremely pure silicon or germanium, which must be free from impurities to better than one part in a billion.

Under certain conditions, an electron that is pair-bonded (with electrons in a neighboring atom) can leave the bond and join the hole in a nearby acceptor impurity. It therefore forms simultaneously a new pair-bond and a new hole. The electron has moved in one direction and the hole has moved in the opposite direction. This concept of holes is very important in transistor operation and can perhaps be further explained by an analogy: suppose a line of cars is waiting at a toll booth at the entrance of a tunnel. As the first car in the line goes through the booth, it leaves a space, or hole, behind it. The second car in line then moves up, leaving a hole behind it. As successive cars move up, the hole they create by their motion moves in a direction opposite to the motion of the cars.

Now let us consider again a piece of silicon that has been doped with aluminum, an acceptor, and upon which a voltage has been impressed. The electrons that are loosened from their pair-bonds move toward the point of positive potential, and the holes move in the opposite direction. To understand the action of a transistor, it is useful to think of these holes as particles of positively charged electricity. An important point is that the hole disappears when an electron fills a vacancy in an acceptor atom. True, another hole is formed, but it is a new hole, the older one having been eliminated.

THE SEMICONDUCTOR DIODE

When n-type germanium or silicon is joined with p-type germanium or silicon—attracted by the opposite charge—the holes in the p-type germanium move toward the junction. At the junction, a region is formed in which there are fewer or no holes nor excess electrons. Meanwhile, in the body of the p-type germanium some holes remain, as well as an excess of negative ions. These ions were formed by adding an electron to a hole. Previously, the charge on the impurity atom was neutral, because the number of electrons in orbit around the aluminum was balanced by an equal number of protons in the nucleus. Once the hole was filled with an additional electron the net charge on the atom became negative, since it carried an extra electron.

On the other side of the junction the arsenic atoms, with five outer electrons, were electrically balanced with five protons in the nucleus. The electrons leaving the outer shell of the arsenic atom leave a net positive charge on it, making it a positive ion. We have, then, what is illustrated in

Figure 68, a p-type region containing holes and an excess of negative ions, and an n-type region, containing some free electrons and an excess of positive ions. There is a field in the junction area caused by the regions of positive and negative ionic charge on either side. The direction of this electric field is such as to prevent additional electrons from crossing to the p-side and additional holes from crossing to the n-side.

The field set up in this region creates a potential barrier at the junction, a voltage—typically a few tenths of a volt—between the n and p regions, even with no external battery connected.

When a battery is connected to a p-n germanium crystal, two interesting effects can occur. When the positive terminal of the battery is connected to the n-side of the crystal, the excess electrons in this region are attracted toward the battery terminal. In the p-side, the holes are attracted toward the negative terminal. The net effect is to create a *depletion region*, as shown in Figure 69. No significant current flows in this depletion region.

If the terminals of the battery are reversed, the effect is entirely different. The negative voltage repels the free electrons in the n-region, moving them toward the barrier and decreasing the size of the depletion region. Action at the positive terminal repels holes (while attracting electrons), similarly reducing the size of the depletion region on the p-side. As the voltage on the battery is increased, excess electrons and holes move into the depletion region. As electrons and holes combine, electrons are drawn across the p-region into the battery. In the p-region, for each hole that an electron fills, an additional hole is formed when an electron is broken loose by the action of the battery. This drift of electrons and holes across the p-n junction and to the battery constitutes an electric current.

Thus, the p-n crystal can be used as a diode, or rectifier of current. If, for example, a source of alternating voltage is connected to it, current will flow only when the p-type material is positive and the n-type is negative. When polarity is reversed, no current flows. The device therefore rectifies current. These semiconductor diodes have pretty much replaced vacuum-tube diodes, which are more costly, cumbersome, and shorter lived than the p-n crystals.

It should be pointed out that joining p and n crystals together is difficult. They cannot simply be cemented together, they must be *grown*, with first one type of material, then the other, produced during the manufacture of the crystal.

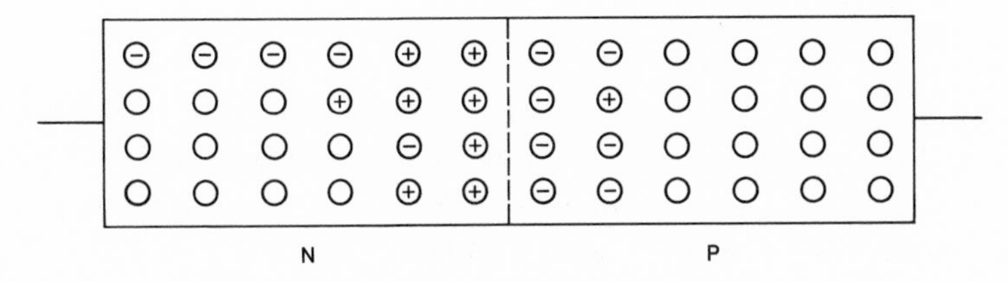

Fig. 68. Representation of the impurity atoms in a semiconductor with a p-n junction. Note that—paradoxically—the n-type material near the junction is slightly positive, the p-type material equally negative.

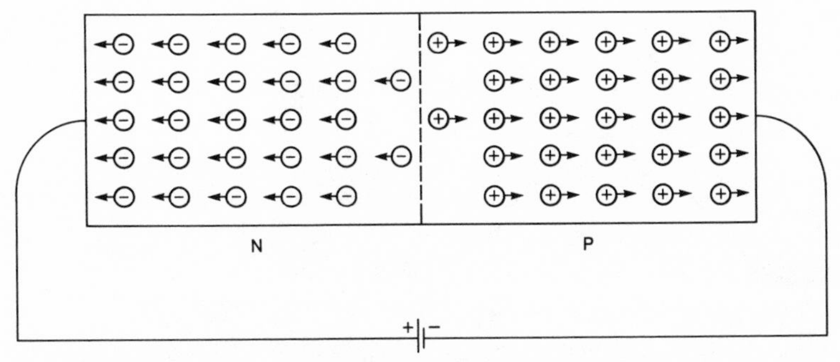

Fig. 69. With polarity shown, the charge carriers (extra electrons and holes) are drawn away from the junction and no current can flow. The junction barrier is increased by the voltage of the battery.

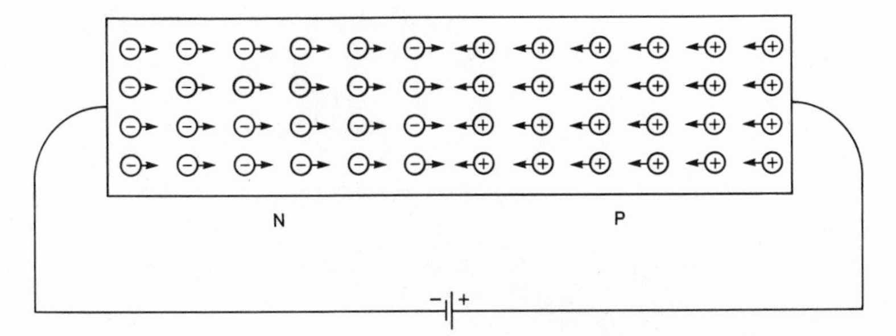

Fig. 70. With this connection, the applied voltage drives the charge carriers toward the junction, where holes and electrons combine. The electrons continue toward the positive terminal and the "holes" toward the negative, creating a flow of electric current.

HOW A TRANSISTOR WORKS

If the crystal consisting of p and n material is expanded into three sections, p-n-p, or n-p-n, an entirely different device is formed. In practice, the center section is thin as compared with the outer two. The devices are illustrated in Figure 72. As shown, the left junction is forward-biased and the right reverse-biased. If batteries are connected as shown, holes are attracted into the center section. If this center section is small enough (and in reality it is no thicker than a human hair), the holes move through the center section into the depletion region at the n-p junction to the right, through it, and into the p-region, attracted by the negative ions there. Because most of the holes pass right through the thin center section, few

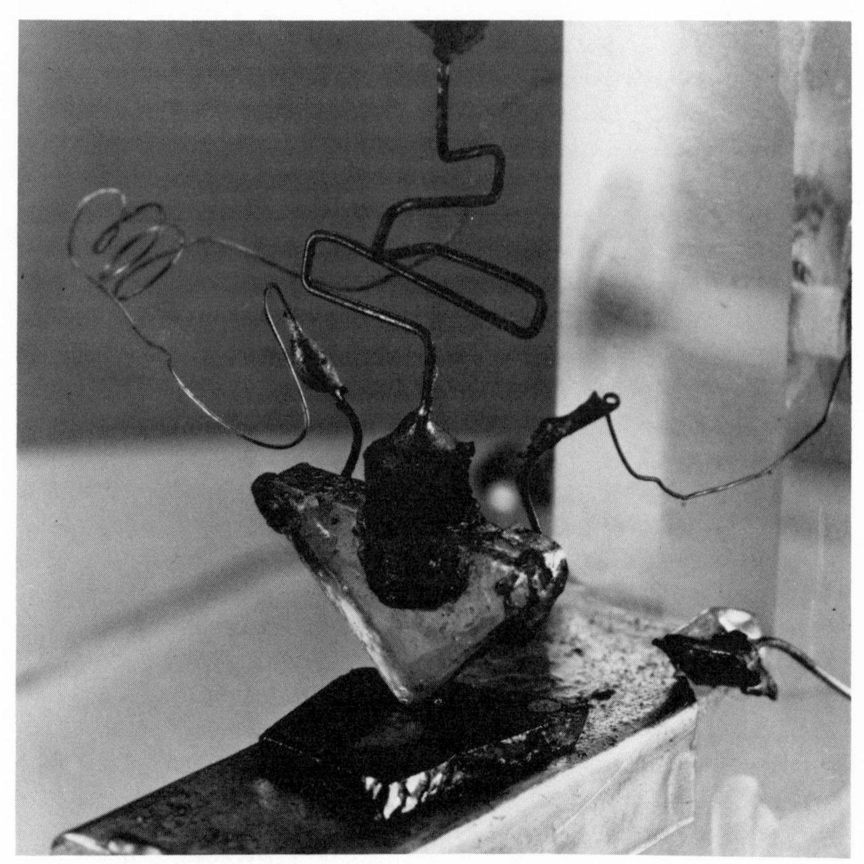

Fig. 71. The world's first transistor. *Courtesy of Bell Telephone Laboratories.*

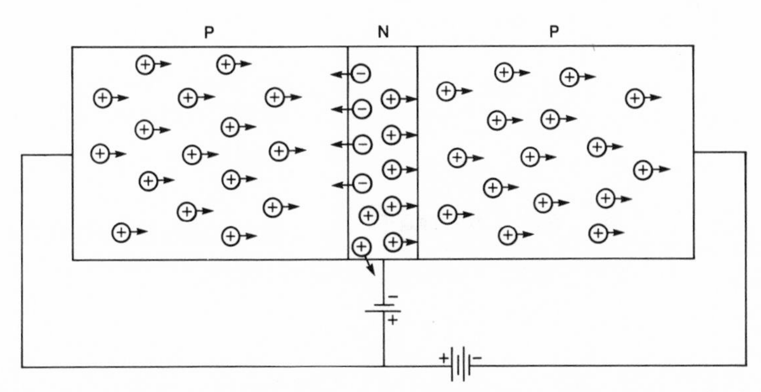

Fig. 72. P-n-p transistor action. For an n-p-n transistor, simply reverse the battery potentials and change the (+) holes to (−) electrons.

electrons flow into it from the battery, and fewer electrons are drawn in to fill these holes. This is due to the thinness of the center (base) section, and the fact that the p-region on the left is more heavily p-doped than the center region is n-doped. The left section of the crystal is called the *emitter*, the center section the base, and the right section, the *collector*.

In the common-base circuit of Figure 72, transistor action occurs because a small emitter-base voltage causes holes to cross the forward-biased p-n junction and enter the base. These are swept across the thin, lightly doped base with little loss from recombination and continue across the reversed-biased base-collector junction, aided by the negative voltage ahead of them.

The voltage across the (reverse-biased) base-collector junction can be high; the voltage required to create the current at the emitter-base junction (forward-biased) is always low (less than 1 volt). The input power is therefore far less than the power that can be delivered across a resistance connected externally in the base-collector circuit, since the current is the same across the two junctions. Hence, large output power is controlled by small input power, and gain is obtained. The action is somewhat like that of a vacuum-tube amplifier, in which a small change of grid voltage can produce a large change in plate current.

Figure 72 shows a p-n-p transistor. The action is identical in an n-p-n transistor—to illustrate n-p-n action simply reverse the battery voltages and change the (+)(−) marked holes to negative electrons.

17

Signals in Space

It was 4:17 P.M. EST, Sunday, July 20, 1969, and the words, profoundly moving in their simplicity, stirred countless millions of people throughout the world: "Houston: Tranquility Base here. The Eagle has landed." Thus was mankind informed that the Lunar Module, carrying Neil Armstrong and Edwin Aldrin, had landed on the desolate plain of the Sea of Tranquility.

Some six and a half hours afterward, people all over the earth watched breathlessly as Armstrong labored down the ladder and carefully implanted the first footprint on the moon. One of man's great achievements was viewed live by hundreds of millions of people as it was happening a quarter of a million miles from the Earth. A communications miracle had occurred.

The greatest feat in the history of technology demanded the most far-reaching and complex collection of communication systems ever applied to a single project. On the ground, it was necessary to establish a network of transmitting and receiving stations to keep in contact with the astronauts while they were on the moon, in the Lunar Module, and in the Command Module. The Earth networks consisted of thirty-nine giant computers, fourteen ground stations, four ship stations, and eight aircraft, each equipped with antennas ranging from just a few feet in diameter to the massive 210-foot parabolic dish antennas at the Earth stations in Goldstone, California, and Parkes, Australia. Depending on the Earth station receiving the signals from the moon, the data was relayed by land line, submarine cable, communications satellite, or by shortwave radio to Mission Control Headquarters in Houston. From Houston, the TV signals were distributed to world television networks via land line, cable, and satellite once again.

But the television signals were just a small part of the communications systems that made the moon mission possible and successful. The systems in operation that were crucial to the success of the mission were telemetry, tracking, and voice communication.

TELEMETRY

Telemetry is the process of measuring data at one point and transmitting it to a distant point for evaluation and action. Information about the astronauts' physical functions—pulse rate, heartbeat, and blood pressure—was transmitted by telemetry. Conditions aboard the spacecraft itself, including temperature, pressure, and radiation, were also telemetered, in addition to data from the onboard computers. All measurements were made by sensors that converted the information to electrical impulses transmitted at the rate of hundreds of data units per second to the Earth stations below. The instruments aboard a spacecraft measure data, then convert it to electrical impulses by *transducers;* the signals are then fed into a *coding box,* which modulates the carrier wave that beams the telemetry data to Earth.

The instant a spacecraft leaves its launch pad, telemetry is the primary means of obtaining data about the condition of the astronauts and their spacecraft. At liftoff, the craft is guided with the help of telemetry. The trajectory of the spacecraft can be regulated by impulses sent by radio from the Mission Control Center. The structural, thermal, power supplies, control equipment, and radio systems are monitored by radio also.

TRACKING

Information about the location of a spacecraft at any instant is vital to the scientists in charge of a space mission. An error of just a few seconds of angle can cause the tragic failure of an operation that took years to plan. Consequently, the precise location, direction, speed, and trajectory of a space vehicle must be known at all times. The Apollo 11 craft was tracked by angle measurements—using an interferometer, which receives signals from the spacecraft as measured at two different points on Earth and compares the angle of arrival of these signals as well as the time difference in their reception, to determine position. A second, more sophisticated system, developed during the 1960s, employed the most advanced Doppler radars in the microwave region of the electromagnetic spectrum.

Each of the spacecraft in the mission—the Command Module, the Service Module, and the Lunar Module—was also equipped with radars for tracking, acquisition, and rendezvous. The Lunar Module was also equipped with radars that fed the onboard computers information about the altitude above the lunar surface and the velocity of the craft during landing on the moon and liftoff.

VOICE COMMUNICATION

The communication networks of Apollo 11 required that astronauts on the surface of the moon be able to communicate with each other by voice as well as with the Lunar Module, the Command Module, and Earth. Systems also linked the Lunar Module with Earth, the Command Module with Earth, and the Lunar Module with the Command Module. All these systems were built with redundancy—if one failed, a second was available for service. Absolutely reliable communication was one essential of a successful mission.

Voice communication between Armstrong and Aldrin, and between the astronauts on the moon and the Command Module, was carried on with amplitude-modulated VHF transmissions. The backpack transmitters used by the lunar astronauts had a power of ½ watt, too weak to be heard on Earth. Impulses were relayed by the Command Module on frequencies in the UHF portion of the spectrum with a higher powered transmitter. Twelve antenna systems were used on the moon and in lunar orbit to transmit information to Earth.

The successful mission to the moon culminated a quarter-century of progress in science that is probably unparalleled in human history. The period from 1944 to 1969 saw the development of atomic energy, the transistor, solar cells, miniature electronics, rocket power, space science, the maser and laser, and the utilization of the microwave region of the spectrum. Scientific progress during this golden quarter-century of science will influence the course of human events for generations to come.

COMMUNICATION SATELLITES

Although not as dramatic as manned flight to the moon, a parallel development in space science has had an even more profound impact on human progress and development, as well as the course of human history. The development of Earth satellites for communications purposes has visually and aurally linked many parts of the globe so that hundreds of millions of people on Earth can simultaneously view an event as it is occurring. By lowering the communications barriers that have kept people apart for centuries, an increased awareness of man's relation to man has resulted.

Until World War II all man's efforts at communication between different locations on Earth, whether along the ground, under the sea, or via the upper atmosphere, had been in regions relatively close to the Earth's surface. Shortly after World War II it became apparent that the frequency bands available for long-distance communication were becoming crowded, and that a severe shortage of frequency space for such communication purposes would be inevitable in the near future.

Frequencies above 50 MHz, which showed promise because of their greater information-carrying capacity, had limited usefulness for long-range communication, because the ionosphere was porous, for the most part, to frequencies above 25 to 30 MHz. A new communications technology was called for, as well as new studies of propagation of radio waves through space.

In the October 1945 issue of *Wireless World*, Arthur Clarke, scientist and science-fiction writer, wrote an article, "Extra-Terrestrial Relays," in which he proposed artificial Earth satellites equipped with transmitters and receivers orbiting high above Earth and relaying messages between remote locations on the surface of the globe.

On January 11, 1946, the first extraterrestrial communication in history took place. Under the code name Project Diana, named for the ancient moon goddess, U.S. Army Signal Corps engineers bounced 112-MHz radar signals off the moon and detected them on Earth, indicating not only that these frequency ranges were potentially useful for transmitting over great distances through the Earth's atmosphere, but also that this could be done with relatively low power, an important consideration at the time.

In 1954 John R. Pierce of Bell Laboratories, in an address before the Princeton, New Jersey, section of the Institute of Radio Engineers, outlined the technical feasibility of communications satellites, using transistorized communication circuits. No Earth satellite had yet been launched, and although some greeted Pierce with skepticism, Bell Laboratories scientists remained interested. All the required technology was already at hand—microwave technology, much of it already in use in microwave communication across the United States, transistorized circuitry, rocket propulsion—it was all there, waiting to be assembled.

Whatever doubts remained about the feasibility of communications satellites were quickly dispelled when Sputnik I was launched in October 1957 and Explorer I several months later. The existence of Earth satellites accelerated the program for developing communications satellites. Within one year of Explorer I, experimental Project Score, the first communications satellite, was launched in December 1958. The payload consisted of an 8-watt transmitter, a receiver, and a tape recorder. The battery-powered repeater was capable of carrying a single voice channel or seven teletype channels either in real time or by tape recording in the satellite for delayed transmission to any of the four Earth stations. The spacecraft's receivers were tuned to 150 MHz, and transmission from Score was on 132 MHz. The chemical batteries powering the Score satellite failed within a month of launch, but the impact of the project was lasting, for among the voices it carried was that of President Eisenhower, who recorded a Christmas message for all the people of the world.

While the Score operation was short lived, it demonstrated the feasibility of active communications satellites: those that could receive information and retransmit it with an on-board transmitter. The early stages of the space communications program, however, consisted of passive satellite tests—that is, the use of objects in space as *reflectors*, rather than *relayers* of waves. The first tests using passive satellites had been conducted in 1946, when radar signals were bounced off the moon. In late 1960, another experiment to test the feasibility of long-range communication by reflection from passive satellite surfaces was undertaken, with the launch of Echo I.

Echo I. This satellite, a large inflatable sphere of Mylar polyester plastic coated with vapor-deposited aluminum, was launched during the early morning hours of August 12, 1960. That night, the first manmade passive communications satellite moved brightly and impressively across the evening sky.

The satellite, which was developed at NASA's Langley Research Center, had a diameter of 100 feet when fully inflated and a surface area of some 32,000 square feet. Its thickness, 0.0005 inches, was about half that of the cellophane on a package of cigarettes.

Because of its aluminum coating, the satellite reflected up to 98 percent of the radio energy beamed toward it, at frequencies up to 20 GHz.

Echo I was carried into orbit like an accordion inside a magnesium container. Once in orbit, the deflated sphere was ejected from its container, whereupon it began to inflate as small quantities of air within it started to expand. Ten pounds of benzoic acid and 20 pounds of antroquinone, sublimating powders which pass from the solid to the gaseous state without liquifying in the near-vacuum of outer space, inflated it further.

Since it was expected that the sphere, once inflated, would receive many small punctures from micrometeorites, gases with very low vapor pressure were used. Low internal pressure slowed the leakage of gas from the satellite and helped it retain a near-spherical shape for about a year before it began to distort noticeably.

Unlike previous satellites, Echo I carried no instruments to gather scientific information and return it to Earth. On its surface were two wafer-thin beacon transmitters operating on 107.94 MHz and powered by seventy solar cells supplying energy to five nickel-cadmium storage batteries generating 10 milliwatts of power. The total weight of the satellite was 167.4 pounds.

The principal purpose of the Echo I experiment was to determine whether it was feasible to use passive reflectors in orbit to transmit radio and television signals over long distances. The results indicated that it was indeed possible to communicate over long distances with passive reflectors. Echo I reflected hundreds of transmissions, teletype signals, facsimile photographs, and two-way telephone conversations, as well as transcontinental and transatlantic signal relays.

Most of the Echo I experiments were conducted between the Bell Laboratories station at Holmdel, New Jersey, and the NASA Jet Propulsion Laboratories at Goldstone, California. The Holmdel transmitter operated at a frequency of 960 MHz, while Goldstone operated at 2,390 MHz.

Both stations used large dish antennas to transmit. Goldstone also used a dish to receive, but Holmdel used a newer horn reflector, which is shaped somewhat like a steam shovel, and is designed to reduce noise levels.

Two hours after Echo I was launched, the following message was transmitted from Goldstone to Holmdel and received clearly:

> This is President Eisenhower speaking. It is a great personal satisfaction to participate in this first experiment in communications involving the use of the satellite balloon known as Echo. This is one more significant step in the United States program of space research and exploration. The program is being carried forward vigorously by the United States for peaceful purposes for the benefit of all mankind. The satellite balloon which has reflected these words may be used freely by any nation for similar experiments in its own interest. Information necessary to prepare for such participation was widely distributed some weeks ago.

By August 12, 1963, Echo I had been in orbit three years. In that time it had traveled over 417 million miles and made 13,555 orbits. Due to constant bombardment by tiny meteoroids its skin had become heavily wrinkled, and although it retained its general spherical shape it was no longer useful in communications experiments.

Project Echo provided much useful information about passive communications satellites, and proved without question that they would work. But the feeling among planners was that active satellites provided the real key to the future. The question was, should they be of the *nonsynchronous* or *synchronous* type?

The nonsynchronous system uses a relatively large number of satellites in orbits at varying altitudes, ranging from several hundred to several thousand miles. In such a system, the satellites are placed in certain nominal orbits with little consideration to maintaining the position of one satellite relative to all the others. As a result, to determine the probability of having any one satellite in view of two particular Earth stations, the satellites must be regarded as being randomly situated.

The second system under consideration required the orbiting of a limited number of satellites in "stationary orbits." Because their periods would be exactly equal to the period of rotation of the Earth they would appear to be stationary above a particular point on the Earth's surface. Such satellites would have an altitude of approximately 22,300 miles.

The nonsynchronous and synchronous systems differed primarily in the requirements of the satellite and the Earth station. For example, to put a satellite into a random orbit required simpler satellites and relatively simple launch vehicles. On the other hand, such a system required rather sophisticated Earth stations with elaborate antennas and tracking facilities to follow the satellites.

Stationary satellites in a synchronous system are inherently more complicated because of the rigid requirements of altitude and "attitude" control. (Its antennas must face the Earth at all times.) The Earth station, on the other hand, would be relatively simple, the antennas needing only to follow minor apparent movement of the satellites resulting from gravitational effects and the deviations in motion resulting from the difference between the orbit of the satellite and the actual shape of the Earth.

Courier 1-B. This satellite, launched in October 1960, was designed exclusively for military use. It tested the feasibility of employing nonsynchronous active repeater satellites in global communications networks.

The total payload of 500 pounds included 300 pounds of electronic equipment, including 1,300 transistors and 19,200 solar cells. The communications system consisted of thirty-eight major components, among which were four microwave FM receivers, four microwave FM transmitters of 6 watts output each, two VHF transmitters for telemetering purposes, a 50-mW transistorized VHF beacon transmitter to provide tracking signals to Earth stations, and five tape recorders. In addition, two microwave and four whip antennas extended from the satellite's equatorial belt.

The Courier 1-B satellite was the greatest step forward in space communications to that date and proved beyond a doubt the feasibility of active satellite communications systems. Since it was designed for military use, the public did not participate in the relay of traffic, which consisted both of active and delayed repeats of teletype, voice, and facsimile. To maintain privacy, the craft contained a command decoder which told the Courier when to record and when to transmit its messages. The satellite operated successfully for about three weeks, when an undetermined malfunction put it out of action permanently.

Telstar. "This is Fred Kappel talking, calling from the Earth station, Andover. . . . The call is being relayed from our Telstar satellite. . . . How do you hear me?"

At a telephone in Washington, D.C., over 400 miles away, Lyndon Johnson, then Vice-President of the United States, replied, "You are coming through nicely, Mr. Kappel." The first telephone conversation via an orbiting active communications satellite had begun.

During the hours immediately following the historic launch of July 10, 1962, many communications firsts were accomplished: the first space telecast, the first transatlantic telecast, the first two-way transatlantic voice and video transmissions via an orbiting satellite. Telstar had begun a new chapter in the history of long-distance communications. During the weeks that followed, live and taped transatlantic TV in black-and-white and color, teletype, facsimile, and high-speed data were transmitted regularly via Telstar.

As an illustration of the awesome potential of active broadband communications satellites, in October 1962 a series of tests were conducted to determine whether American Telephone and Telegraph Company's Telstar could work with data-processing equipment. During one nineteen-minute test of data transmission from France, almost a billion units of information were transmitted with only one error. This is equivalent to almost 1.5

Fig. 73. Telstar II being checked out before launch in 1963. *Courtesy of NASA.*

million words per minute, and is faster than 18,000 stenographers typing at the same time.

The data channel that Telstar provided had many times the capacity of other transatlantic data-transmission facilities. The messages sent included medical data, bookkeeping data, computer-input information, and machine-tool programming instructions. The experiment, which spanned nearly 6,000 miles, was conducted from computer memory to computer memory, as well as from computer to magnetic tape. The intelligence transmitted in one minute of typical tests was equivalent, in number of words, to about three times the number of alphabetically listed words in an unabridged dictionary, or fifteen novels, or about 230 hours of speech at normal speed.

In addition to making one of the most significant breakthroughs in the history of international communications, Telstar provided invaluable data on the effects of radiation on space-communications equipment, on tracking techniques, and on internal and external stresses on the communications satellite in space environment.

Technical Data. The Telstar communications experiment was intended primarily to test whether a satellite could receive radio signals from the ground, amplify them ten billion times, and retransmit them on another frequency. Because of its great height above the Earth, the satellite, which was essentially a broadband microwave relay station, was able to transmit these signals over very long distances.

Telstar was roughly spherical, with a diameter of 34.5 inches. It consisted of seventy-two flat faces, sixty of which were covered with 3,600 solar cells to supply energy to nineteen nickel-cadmium batteries. Mirrors were placed on three of the facets to reflect sunlight to facilitate optical tracking. Two receiving and transmitting antennas girdled the satellite about its equator. These radiated tracking signals as well as those of the communications experiment. A helix at the top of Telstar functioned in the telemetry, command, and the continuous beacon circuits.

The solar cells and batteries, which converted sunlight into electricity, supplied about 15 watts during parts of the early orbits when the satellite was in sunlight. Afterward, due to radiation effects, this output diminished steadily.

The electronic equipment was suspended inside an aluminum canister with nylon cord to help absorb shock and vibration. It consisted of 2,528 semiconductor devices, of which 1,064 were transistors and 1,464 diodes.

Broadband Communications Circuit. Signals were beamed to Telstar on a frequency of 6,390 MHz and were received by the satellite at a level of 2.5 microwatts. This weak signal was transformed to 90 MHz, a frequency that could be handled by reliable, long-life transistors. The 90-MHz signal was amplified about one million times by fourteen germanium transistors.

The amplified signal was then mixed with another crystal-controlled oscillator signal so that the resulting mixture was centered on 4,170 MHz, the satellite's transmitting frequency. This signal was amplified once more by the only electron tube in Telstar—a foot-long, pencil-thin traveling-wave glass tube. This was the only device of such size and weight capable of high amplification of such a broadband signal.

This same traveling-wave tube also amplified the 4,080-MHz signal that was transmitted at about 0.02 watts to serve as a beacon for precision tracking by ground stations.

Telemetry. In all, 115 measurements were made and reported on by the Telstar satellite. Although the communications aspects of the program were the important ones, the measurement of environmental conditions was also significant. It was necessary to find out whether radiation in space was such that the life of a communications satellite, operating perfectly at first, would be long enough to make active communications satellites financially practical.

Among the more important measurements made were the density and energy levels of free protons and electrons within the spacecraft, temperature on the skin of the satellite and inside the chassis, pressure inside Telstar, the currents and voltages in several dozens of electronic components in critical circuits, and the amount of sunlight on several points on the skin of the craft.

The information was transmitted on a frequency of 136 MHz at a power of 0.25 watts. The transmitter was in constant operation, even when no information was being transmitted, and served as a radio beacon to assist in the tracking.

The satellite transmitted its measurements on command from the ground (over a separate command circuit) and continued transmitting data until a command to stop was given.

Command. When all Telstar's circuits were in operation the power drain from the battery was greater than the average rate at which its solar batteries could replenish the power supply. To conserve power, it was necessary to turn off many of the vehicle's experiments when not needed. The equipment that did that work was the *command system.*

The command system of Telstar I included two receivers for pulse-coded commands from the ground on a frequency of 120 MHz. The system also included two decoders to translate the pulses into usable instructions, and a network of electrically operated switches, which turned the appropriate circuits on and off.

The vital importance of the command system demanded that receivers and decoders be installed in pairs to ensure that these critical operations continue to function if one of the devices should fail. This duplicate equipment was the only redundancy on the craft.

Telstar I was designed so that the 136-MHz telemetry signal would continue to radiate up to two years, after which a timing device automatically stopped operation. The principal frequency of 4,170 MHz, as well as its beacon companion of 4,080 MHz, could be cut off by command. If the command circuits failed, the power drain was such that operation on these frequencies would end in several hours.

Telemetry and command accounted for 93 percent of the semiconductor devices used on the satellite. In all, 2,354 semiconductors out of the 2,528 for the entire satellite were used in these circuits. They took such a large proportion of the weight and complexity of the satellite because of the experimental nature of the program.

Ground Stations. The main Earth station for carrying out active communications satellite experiments was at Andover, Maine, about 80 miles northwest of Portland. The site is surrounded by mountains, which help shield the antenna from unwanted outside microwave and UHF noise. The station is equipped to track the satellite, compute its orbit, send commands, receive telemetered information, and conduct all the major communications experiments for which the satellite was designed.

To receive very weak signals (as low as a billionth of a watt) it was necessary to build a very large antenna. It was also necessary to screen the sensitive element of the antenna from the unwanted radiation emitted by trees, the Earth, and even people. To add to the difficulty, the system had to be able to track a 34-inch object moving through space at about 16,000 miles per hour at a distance of thousands of miles.

To solve the problem, a 380-ton steel and aluminum rotating structure with an overall length of 177 feet was built. It carried two houses that contained the receiving and transmitting equipment. The antenna itself, the largest horn yet built, had an opening of about 3,600 square feet.

The giant Andover horn has requirements more exacting than any structure of comparable size ever built. It must track its tiny target smoothly and continuously to an accuracy of better than two seconds of arc and yet withstand the stresses of wind and rapid temperature changes.

The horn was covered with an inflated radome, 210 feet in diameter and 161 feet high. This structure protects it from weather and is transparent to radio waves. The antenna which, for its size, is more accurate than a fine watch, is mounted on a 70-foot diameter rotating wheel machined to a tolerance of 0.03 inches.

The permanent radome is made of dacron and synthetic rubber. Only a sixteenth of an inch thick, it would cover three acres if laid out flat, and weighs 20 tons. It is held in place and maintains its rigidity by air pressure of less than 0.1 pound per square inch above atmospheric pressure.

Despite the size and sensitivity of the Andover horn, reception would not be possible without the use of a *maser*, an extremely sensitive amplifying device, cooled by liquid helium to −456° F, as a central element. Transmission is facilitated by one of the largest traveling-wave tubes ever built—a water-cooled device more than 4 feet long. The power output of the 25 MHz bandwidth transmitter was about 2 kW.

The British participated in the Telstar experiment from their Goon-

Fig. 74. The 380-ton horn antenna at Andover, Maine. *Courtesy of COMSAT.*

hilly, England, station, which is equipped with a steerable parabolic antenna about 85 feet in diameter. The station can transmit and receive TV, facsimile, voice, and data transmissions via satellite.

A third participant, the French, have a station at Pleumeur-Bodou in Brittany. It is almost identical to the Andover system, and is equipped to conduct TV, voice, facsimile, and data experiments.

LATER COMMUNICATION SATELLITES

Relay I. Relay I, NASA's next space-communications satellite experiment, was designed and built by RCA's Astro-Electronics division in Princeton, New Jersey. It was intended, among other things, to determine the feasibility of satellite communication between North and South America. The communications and scientific objectives of the Relay satellite were essentially the same as those of Telstar.

The satellite differed from Telstar in its engineering approach to the relay of voice signals. With about the same weight, the Relay satellite carried many more redundant subassemblies than Telstar. It had two command receivers, two command decoders, two telemetry transmitters, two continuous-wave beacons for ground tracking, two wideband transmitters, two wideband receivers, and two regulated power supplies for some of the electronic equipment.

Syncom. The first attempt to place an active communications satellite in a stationary orbit ended in failure during February 1963. Trouble first developed about five hours after launch, when a solid-fuel rocket motor gave the satellite a "kick" in an effort to give it enough energy to take it out of an elliptical orbit and put it into a circular orbit at an altitude of 22,300 miles. Unfortunately, the motor kicked too hard for the radio equipment aboard, and rendered the gear inoperable.

If successful, the satellite would have traveled at approximately 16,000 miles per hour, making it appear to be hovering in a stationary position over the same spot on the Earth's surface. Any two locations to which it was mutually "visible" could be in 24-hour contact via the satellite. Theoretically, three such satellites, appropriately placed, could provide a continuous, worldwide space-communications system.

Fig. 75. Syncom, first synchronous orbiting communications satellite. *Courtesy of NASA.*

Syncom II. One of the resounding successes in space communication took place on July 26, 1963, when a new breed of active space satellites rocketed into orbit from Cape Kennedy. The craft was injected into a near-synchronous orbit with an apogee of 22,632 miles and a perigee of 21,180 miles. It was drifting eastward at a rate of about 7.5 degrees per day. During the day following the launch the first of the in-orbit maneuvers was commanded from a vessel anchored in Lagos harbor, Nigeria. By firing the satellite's hydrogen-peroxide jet system along the spin axis of the satellite, the apogee was made higher and the drift was reduced.

On July 31, the satellite's antenna was aligned so that the radiating beam was always pointed toward Earth. In two series of pulses, Syncom II was turned about 85 degrees. This maneuver also changed its drift rate to nearly 7 degrees per day westward. Three more jet firings were required before the drift was reduced to near zero, so that Syncom II was moving with the Earth.

The orbit finally achieved by Syncom II was near-stationary, as distinguished from a true stationary orbit in perfect alignment with the plane of the equator. Instead, it hovered over Brazil and the Atlantic Ocean while moving in a figure-eight pattern 33 degrees north and south of the equator. In the short span of three weeks, Syncom II, the world's first satellite to operate at synchronous altitude, chalked up a notable series of successes.

Within two weeks after launch, Syncom II demonstrated convincingly the soundness of the synchronous orbit communications satellite concept. In tests conducted by project engineers and technicians, the NASA satellite handled voice, teletype, facsimile, and data transmissions between surface communications stations at Lagos, Nigeria, and Lakehurst, New Jersey.

Two communications milestones were achieved by Syncom II in September 1963. One of these was the first transmission of speech and teletype via a communications satellite to a moving ship at sea. Voice and teletype were transmitted from a ground station at Fort Dix, New Jersey, to the surface terminal ship *Kingsport* while at sea, some 40 miles west of Lagos, Nigeria. The Government of the Federation of Nigeria made an important contribution to the project by permitting the *Kingsport* to be stationed in Lagos harbor and making other facilities available. Communications tests were conducted between a transportable ground station at Lakehurst, New Jersey, and the station aboard ship. Although the *Kingsport* was rolling, pitching, and yawing, the ship's 30-foot antenna remained locked on Syncom II, and the officials who conducted the tests reported the voice transmission to be excellent.

The U.S. army ground station at Fort Dix, New Jersey, also participated in the first TV transmissions relayed via a synchronous orbiting satellite. Due to bandwidth limitations, no audio was transmitted, but officials at the AT&T ground station at Andover, Maine, indicated that reception at that ground station of test patterns sent via Syncom II was of good quality.

Intelsat. On February 1, 1963, the Communications Satellite Corporation, COMSAT, created by an act of Congress, was incorporated in the District of Columbia. The Communications Satellite Act of 1962, to establish a commercial communications system by itself or in cooperation with other administrations, provided that private funds be used.

In August 1964, COMSAT signed an international agreement by which a global communications satellite system, INTELSAT, the International Telecommunications Satellite Consortium, would be established on a joint-venture basis. COMSAT holds the largest individual interest in INTELSAT.

A new communications era was heralded in 1965, when Early Bird, the world's first commercial communications satellite, was successfully launched and placed in synchronous orbit by NASA on June 28, 1965, to serve the communications requirements between North America and Europe. Early Bird had a capacity of 240 two-way telephone circuits, or one television channel. Within two years, INTELSAT had launched three additional satellites to provide essentially worldwide coverage via satellite.

Subsequent satellites have had successively greater information-carrying capacity. The latest of these, the INTELSAT IV series, are the largest commercial communications satellites ever launched; they are each capable of carrying an average number of five to six thousand telephone circuits simultaneously, or 12 color television channels, or a combination of telephone, TV, data, and other forms of communication traffic. All are in synchronous orbits.

SPECIAL PURPOSE SATELLITES

The word *telemetry*, from the Greek *tele*, far off, and *meter*, measure, means measurement at a distance. Telemetered data has been indispensable in furthering our knowledge about the universe, the space environment, and our planet. From an economic point of view, meteorological satellites are second only in importance to the communications satellites. Since 1960, these satellites have returned to Earth literally hundreds of thousands of photos of parts of our planet.

Tiros I was launched on April 1, 1960, as part of an extended Tiros

Fig. 76. Array of helical antennas at Wallops Island, Virginia. It receives data from orbiting weather satellites. *Courtesy of NASA.*

(Television and Infrared Observation Satellite) program to return to Earth television pictures of cloud formations and patterns. They were also designed to transmit scientific data to ground stations for analysis and operational and research use, leading to improved weather analysis and forecasting services.

The drum-shaped Tiros satellites were approximately 20 inches high and 42 inches in diameter. They weighed nearly 280 pounds, and their orbits ranged from 450 to 500 miles above the Earth. Power was supplied by 9,260 solar cells energizing 63 nickel cadmium storage batteries. Five transmitters relayed data from the satellite to ground stations. Each of the two television camera systems had a 2-watt transmitter operating on 235 MHz. One 2-watt 237.8 MHz transmitter relayed infrared experiment data. Two tracking beacons, operating continuously on frequencies in the 136 MHz band, relayed satellite telemetry data such as temperature, pressure, and battery-charge level.

In all, ten Tiros satellites were launched in the highly successful program before the second generation of weather satellites, the Nimbus series, was undertaken in August 1964. Nimbus was originally conceived as a meteorological satellite whose primary mission was to provide atmospheric data for improved weather forecasting. Subsequent spacecraft have added more sophisticated sensing devices, and on each succeeding Nimbus launch the capability and performance of the craft has increased until it now encompasses a wide range of Earth sciences, including oceanography, hydrology, geology, geomorphology, geography, cartography, and agriculture. The most recent satellites in the Nimbus series provide moisture, temperature, and vegetation data in areas frequently covered by clouds because they are equipped with a series of sensors, including an infrared temperature profile radiometer.

In infrared radiometers, lenses focus the infrared radiation on a detector made from a photoconducting material, such as lead selenide, which becomes a good electrical conductor when illuminated by infrared. The amount of current passed by the detector is proportional to the intensity of the infrared and therefore to the amount of heat being radiated from Earth to outer space. (The amount of infrared radiation emitted by an object is proportional to its temperature.)

Infrared radiometers can be made sensitive to various wavelength ranges or channels with filters. For example, the high resolution infrared radiometer on Nimbus I was sensitive to only that radiation between wavelengths of 3.4 and 4.2 microns (millionths of a meter). Radiation of this wavelength is emitted from cloud tops and gave Nimbus I a way of mapping cloud cover at night. An infrared channel between 6 and 7 microns helps determine the amount of absorption caused by water vapor in the air. Data from such a channel helps construct worldwide humidity charts of the upper atmosphere. Analysis of the warm Earth's radiation by radiometers, spectrometers, and other optical equipment can provide atmospheric temperature and humidity profiles, vertical water vapor distribution, vertical ozone distribution, and surface temperature data. When added to photos taken in visible light by weather-satellite cameras, meteorologists can see the world's weather more completely than ever before.

The Pioneer and Mariner series have also supplied important information about the interplanetary environment. Pioneer I, launched in October 1958, sent back the first data pointing to the existence of the Van Allen radiation belt, which girdles the Earth. By the time the Pioneer III spacecraft, launched in 1959, was three months in orbit, it had sent back enough data to verify the existence of a second radiation belt around the Earth. The Van Allen radiation belts are regions beyond the ionosphere, but within the influence of Earth's magnetic field, where particles from the sun are trapped and held. The belts extend from hundreds to tens of thousands of miles into space.

Pioneers X and XI returned television pictures of the planet Jupiter, its moons, and its mysterious giant red spot. Both spacecraft passed through the asteroid belt before passing Jupiter on a course that would take them out of the solar system, the first objects made by man to leave the system and pass into our galaxy.

The Pioneer missions to Jupiter and beyond opened the era of exploration of the outer planets, and were partly intended to develop technology for other outer planet missions. Jupiter is so distant that radio signals from it, traveling with the speed of light, take 45 minutes to reach Earth. Signals transmitted by Pioneer X and XI with their 8-watt transmitters reach the vicinity of Earth with a power of 1/100,000,000,000,000,000 watt. If collected for 19 million years, this power would light a 7.5-watt Christmas tree bulb for one-thousandth of a second. The weakness of these signals strain the facilities of NASA's 210-foot Deep Space Network (DSN) dish antennas.

The communications system provides for two-way communication between Earth and the spacecraft. For reliability, the system is fully redundant. It depends on the sensitivity of the DSN's 64-meter (210-foot) antennas and their receivers which can hear the fantastically weak signals from Pioneer at Jupiter. When used in reverse as transmitters, these 64-meter antennas have such precision and effective radiated power (up to 400 million watts) that outgoing commands are still strong enough to be received when they reach the spacecraft.

The spacecraft system consists of high-gain, medium-gain, and low-gain antennas, used for both sending and receiving. The high-gain antenna is the spacecraft's parabolic reflector dish; the medium-gain antenna is an

Earth-facing horn mounted along with the high-gain antenna feed on struts at the focus of the dish. The low-gain antenna is a spiral, pointed to the rear, designed to provide communication during the few times when the aft end of the spacecraft is pointing toward Earth. Each antenna is always connected to one of the two spacecraft radio receivers, and the two receivers are interchangeable by command, or automatically after a certain period of inactivity, so that if one receiver fails the other can take over.

MARINER II

On January 3, 1963, the first of the planetary explorations by a spacecraft came to a close. On that day contact with Mariner II was lost. It had come within 22,000 miles of Venus and had beamed back the first data from the environs of another planet in the solar system with a 3-watt transmitter operating on 960 MHz. Valuable information about temperature, magnetic fields, and radiation in the vicinity of the planet was sent to Earth.

Subsequent Mariner flights to Venus and Mars furthered man's knowledge of his two closest planets immensely. Mariner IX, in one of the most significant experiments in the entire space program, mapped the entire surface of Mars photographically, making some 500 orbits of the planet to do so.

Mariner Mars missions also supplied data needed to plan and design the next generation of Martian exploratory vessels, the Viking spacecraft. Mariners supplied data on atmospheric composition, atmospheric structure, surface elevations, atmosphere and surface temperatures, topography, and ephemeris information.

The Viking Mars missions were surely among the most spectacular and remarkable experiments in the history of communication. Launched in 1975, Viking I and Viking II arrived in the vicinity of Mars in the summer of 1976, after having traveled approximately 400 million miles.

Each spacecraft was composed of an orbiter and a lander, and once in orbit, they separated. The landers were soft-landed on Mars in late summer of 1976.

While the orbiters provided a complete series of photos of Mars and its satellites, the landers sent back information about the topography, atmospheric composition, climate, and soil composition of the planet. The linear distance from which these data were transmitted exceeded 200 million miles.

Fig. 77. Viking's-eye view of the Martian terrain. Light streak above horizon is a haze, probably of dry-ice crystals. *Courtesy of NASA.*

Fig. 78. Composite photograph of Arsia Mons, one of three large volcanoes. The central depression is about 75 miles across. *Courtesy of NASA.*

It will take years to fully analyze the great mass of information returned to Earth by the Viking satellites.

OTHER PROGRAMS

The use of radio to obtain information that was impossible to attain before 1957 has extended to other areas and other space projects. Among these are the orbiting observatories, Skylab, and many other series that have shed much light about the immediate vicinity of Earth, the ionosphere, and the radiation coming to us from outer space, all of which have added to our understanding of the universe.

18

The Hams in Space

The very nature of amateur radio, "A service of self-training, intercommunication and technical investigation carried on by amateurs . . ." is such that in the seventy years of its existence it has not only kept abreast of developments in the field of communications, but has led the way in advancing new techniques and innovations. From their early major contribution to communications—the development of the shortwaves—amateurs have been pioneers in the VHF and UHF portions of the spectrum, and were among the first to use the moon as a passive reflector of radio waves. Amateurs were among the first to use the vacuum tube in the design of transmitters and receivers; amateur radio was the first service to prohibit completely the use of spark transmission and among the first services to use CW.

In spite of the serious handicaps imposed by a complicated and costly technology, radio amateurs have extended their impressive record of research and innovation into the field of space communications.

OSCAR I

Amateur radio entered the space age on December 12, 1961 (sixty years to the day after Marconi's momentous transatlantic experiment), with the launching of Oscar I, a satellite designed, built, tested, and financed entirely by amateurs. The Project Oscar (Orbiting Satellite Carrying Amateur Radio) Association was formed in 1960 by a group of amateurs whose imagination had been fired by developments in the space sciences during the previous few years. Their objective was to design, build, and test satellites that would operate in the bands allocated to amateur radio.

Fig. 79. Oscar VI, launched October 15, 1972. *Courtesy of ARRL.*

The launch, as ballast aboard a Thor-Agena rocket carrying the Discoverer XXXVI space satellite, was the result of more than a year's frantic activity to build the craft and to obtain permission from the Air Force for the launch. Finally, after a seemingly endless series of exchanges between Oscar Association officials and the Air Force, the go-ahead was given and the history-making Oscar launched from Vandenberg Air Force Base. As the powerful rocket blazed a trail into space a new era in amateur radio began.

Oscar I contained a relatively simple 100-milliwatt beacon transmitter, which carried the greeting "HI" in Morse code to amateurs throughout the world. The transmission rate was proportional to the internal temperature of the craft. The 10-pound spacecraft operated on 144.98 MHz in the amateur 2-meter band for three weeks before reentering Earth's atmosphere. During its lifetime in orbit, amateurs in twenty-eight countries sent over 5,000 reception reports to the Oscar Association. This was a remarkable response, considering that the majority of the world's amateurs operate in the shortwave portion of the radio spectrum, with relatively few equipped to operate on 2 meters.

Oscar II was launched in June 1962. It was virtually identical to Oscar I. It operated eighteen days before being consumed in a ball of flame while returning into Earth's atmosphere.

The first two Oscar satellites, primitive by today's standards, are remarkable in that they were built from spare parts by men who worked evenings and weekends in home workshops, in basements, in garages, in an attempt to contribute to space technology and to further the amateur cause. They not only introduced amateur radio to the space age, but contributed valuable VHF propagation data, and gave the amateur community valuable tracking experience for the adventures that were still to come.

Oscar III, launched March 9, 1965, made telecommunications history. It rode piggyback into space on an Air Force satellite, and shortly after ejection its beacon telemetry transmitter, operating on 145.85 MHz, began sending the signals that informed amateurs throughout the world that it was in orbit. But Oscar III was equipped with more than a beacon transmitter. It was the first active amateur satellite and was able to receive 2-meter signals, amplify them, translate them to another frequency in the same band, and transmit them back to Earth. The 1-watt transmitter remained in operation for two weeks, during which time scores of records were broken, reminiscent of the hectic, furious days of the middle 1920s when shortwave contacts of history-making proportions were made on an almost daily basis.

For the first time transcontinental and transatlantic radio communication in the 2-meter amateur band took place. Oscar III was the *first* free-access active communications satellite ever launched, and during its brief lifetime more than one hundred stations in sixteen countries were able to communicate in a frequency band that had heretofore been used by amateurs to contact their neighbors across town, or at best over a distance of only a few miles. Part of the remarkable aspect of the Oscar III operation was the realization that communication via space satellite was possible without high-power, high-gain, highly sophisticated hardware. Furthermore, Oscar III was launched one month before Early Bird, the first commercial communications satellite orbited by INTELSAT.

Oscar IV was to have been even more revolutionary than its predecessor, with a receiving capability in the 2-meter band, and a transmitting capability in the 70-centimeter band (432 MHz) in the UHF portion of the spectrum. The 3-watt repeater functioned well, but through an unfortunate quirk the satellite was put into the wrong orbit, and only a handful of contacts were made. Of historic importance, however, was the contact between K2GUN in the United States, and UP2ON, in the Soviet Union, the first time direct satellite communication between these two countries had ever been made.

Demonstrating the international nature of amateur communications, Oscar V was built entirely by students at the University of Melbourne under the guidance of the Wireless Institute of Australia. It was the first in the series of Oscar satellites to carry a beacon transmitter operating in the shortwave portion of the spectrum, 29.45 MHz. Detailed observations by amateurs throughout the world indicated several unusual propagation anomalies, including over-the-horizon reception and very long-distance propagation; the satellite signal would appear while the satellite was passing over a point on the opposite side of the Earth from the observer. Simultaneous observations of general propagation conditions revealed substantial dropouts of signal along the path, only over shorter circuits, indicating that the observed signals were probably propagating as the result of *ducting,* that is, the signal traveled between or within ionospheric layers for a substantial distance.

Oscar V was also the first in the series to be actively controlled by ground command. The 29.45-MHz beacon was turned off on weekdays to conserve power in the chemical batteries. Another first for Oscar V was its Magnetic Attitude Stabilization System (MASS), consisting of a simple bar magnet and an eddy current damper which helped stabilize the attitude of the craft along Earth's magnetic lines of force, the way a compass needle aligns itself.

Launched by NASA in January 1970 with a Tiros meteorological satellite, Oscar V generated considerable enthusiasm among hams throughout the world; in addition to the firsts for the satellite, monitoring it throughout the world familiarized amateurs who did not have VHF and UHF receiving equipment with some space techniques, and gave them practice in tracking in preparation for future missions.

Oscar V, which remained active for a month and a half, was the first satellite launched under the auspices of Amsat, the Radio Amateur Satellite Corporation, an international amateur organization based in Washington, D.C., with members from some twenty-five countries. It is made up of individuals as well as groups, clubs, and organizations whose interests lie in the field of amateur space communications activities. Among the projects envisioned during the formation of Amsat was the launch of long-life wideband repeater satellites; active repeater satellites in synchronous orbit relaying AM, FM, single-side-band transmission; and even television.

The first of Amsat's objectives was realized in late 1972, with the launch of Oscar VI, the product of a three-year effort by American, Australian, and German amateur groups. It was a space applications satellite designed to conduct an experimental program of multiple-access communications with a large number of relatively low-powered Earth stations operated by radio amateurs.

Oscar VI was powered by solar cells and battery and weighed some 40 pounds. As its primary payload it carried a 2- to 10-meter linear translator with a bandwidth of 100 kHz. (A linear translator is a device that accepts a transmission in one band, and repeats it in another band. For example, if the satellite accepts a signal at 145.95 MHz, it may repeat the same signal at 29.5 MHz. An incoming signal at 145.955 MHz repeats the same signal at 29.505 MHz, etc.)

The satellite was designed to repeat the strongest signals it received and was capable of relaying up to ten simultaneous signals spaced throughout its 100-kHz operational bandwidth. Orbit information was carried in advance in amateur publications, and on the air by amateur stations manned by the American Radio Relay League.

The input frequency was centered on 145.95 MHz, and the output at 29.5 MHz. Peak power output of the satellite transmitter was 1 watt, and the payload also included a message storage device loadable from the ground with Morse code traffic that could then be repeated by the satellite. Control was through a 21-function command system under the control of Amsat.

Results during the first six months of Oscar VI's operation were spectacular, and many amateur distance records were broken. Communication between the United States and Japan, and the United States and Soviet amateurs, was reported several times, and distance contacts to 4,500 miles often took place. The first instance of satellite communication by an amateur operating a low-power mobile station from a car occurred several months after the Oscar VI launch.

Sent aloft with NASA's launch of a National Oceanic and Atmospheric Administration (NOAA) weather satellite, Oscar VI was a significant step forward in the evolution of amateur radio. Satellite communication has thus broadened the scope of amateur radio, which, with the world as its classroom, has historically served as a training ground in the field of electronics. In the past, amateur radio has performed services in times of emergency which could not be duplicated elsewhere. On many occasions, for example, amateur radio has provided the only source of communication with areas struck by disaster. The 1964 earthquake in Alaska and the June 1970 earthquake in Peru are two examples. In less publicized instances, amateur communication has located urgently needed drugs and sera and has even given medical instructions during emergency surgery. Amateur space communication has broadened the horizons within which ham radio can serve. The Oscar satellites have made it possible to communicate with isolated areas irrespective of ionospheric conditions. The future of amateur radio, dynamic and innovative in the past, appears to promise at least as much in the years to come.

19

Signals from the Stars

In 1931 a young Bell Telephone Laboratories engineer, Karl Jansky, noted a steady, hissing sound from his receiver, which he was using to study the different types of static that affected long-distance telephone communication in the shortwave portion of the spectrum. Atmospheric noise, or static, had been a source of difficulty to the Bell system because any kind of noise degrades the reliability of communications. Jansky, who was chronically ill, requested that he be assigned work that would not exert undue pressure on him, and was given the noise project.

By late 1932, Jansky had traced much of the unwanted noise to thunderstorm activity. But in a paper, "Directional Studies of Atmospherics at High Frequencies," published in the December 1932 issue of the Proceedings of the Institute of Radio Engineers, he described another type of noise: "A steady hiss static, the origin of which is not known." This unexplained noise troubled him, and he concentrated his efforts on tracing it.

Jansky had collected thousands of observations, primarily in the 14-meter (approximately 21 MHz) band. After checking and rechecking the angle of arrival of the strange noise source and consulting with an astronomer colleague, he concluded that the noise he was observing was coming from the Milky Way. One of the important reasons for believing that the signals were coming from outside the solar system was that the recorded noise level from his receiver went through maxima and minima every twenty-three hours and fifty-six minutes, which is the period of rotation of the stars around the Earth.

Jansky published papers in scientific journals describing the phenomenon, and as a result of another IRE paper, "Radio Waves from Outside the Solar System," published in 1933, he made the front page of the *New*

Fig. 80. Karl Jansky, father of radio astronomy, with the rotating antenna he used to discover radio waves from outer space. *Courtesy of Bell Telephone Laboratories.*

York Times. But, strangely, the discovery did not fire the imagination of other scientists and astronomers, who apparently did not recognize the full significance of Jansky's work.

By 1934, Jansky was making observations of extraterrestrial noise in the 4- to 20-meter region of the spectrum, in both the shortwave and VHF bands. He continued to publish his findings, but by 1938, his work shifted to the measurement of the angles of arrival of shortwaves propagating over the Atlantic Ocean. To have continued his radio-noise efforts, it would have been necessary for him to have a large antenna, vertically and horizontally steerable, and such systems were unknown at the time. With the coming of World War II, Jansky shifted his efforts into the rapidly developing field of radar, and after the war he specialized in microwave repeater technology.

It was a radio amateur, Grote Reber, fascinated by Jansky's original papers, who in 1937 took up where Jansky left off. A radio repairman by day, Reber built a 31-foot dish antenna in the backyard of his Wheaton, Illinois, home. He built the antenna himself because he had been quoted a price of $7,000 for the job and could ill afford such a sum. Using his antenna and working at night, Reber made the first radio map of the sky, plotting noise sources in the VHF portion of the spectrum in the region between 150 and 300 MHz. Reber reported radiation coming not only from the Milky Way, but from other "hot spots" in the vicinity of Cygnus, Aquila, and Canis Major, as well as others. Again, there was little enthusiasm among scientists.

In February of 1942 it was accidentally discovered that the sun was an

Fig. 81. Karl Jansky, with some of his original data, describes his observations of extraterrestrial noise. *Courtesy of Bell Telephone Laboratories.*

Fig. 82. World's first parabolic radio telescope, built by Grote Reber in 1937. *Courtesy of the National Radio Astronomy Observatory.*

emitter of radio radiation in the 300-MHz region. The phenomenon was at first misinterpreted because it had resulted in an almost total blackout of radars operating in the United Kingdom, and it was feared that the Germans had developed a highly effective radar-jamming network.

A British physicist investigated and learned that the sun, and not the

Germans, had jammed the radars. Subsequent wartime investigation disclosed that radars in other countries were having similar difficulties in frequency ranges that extended well into the UHF portion of the spectrum at centimeter wavelengths. The Americans had been right. Signals were coming from the stars!

After the war scientists began to take radio astronomy more seriously, curious to learn about the strange "noise storms" that had blocked the military radars. Radio astronomy observatories were constructed in England, Holland, and Australia.

Two early discoveries fueled the rapid expansion of radio astronomy. J.S. Hey, the British physicist who had explained the mysterious radar jamming, observed strange signals coming from an apparently empty space in the constellation Cygnus. Other radio astronomy observatories looked at the mysterious noise source, and slowly pinpointed its location. Optical observations yielded no clue to the unexplained signals, which appeared to pulsate, as if they were twinkling. Four years after Hey's first observation, Walter Baade and Rudolph Minkowski studied the source through the giant Palomar telescope and made a remarkable discovery. The source of radio noise observed since 1947 appeared to be two giant galaxies, some 280 million light-years distant, in collision. Tens of millions of stars rushing at each other at this vast distance had been emitting the mysterious radiation observed by the radio telescopes.

A second startling discovery was made when an Australian, John Bolton, turned his radio telescope to the spot at which Chinese historians reported a giant explosion in the sky in 1054 AD. So gigantic had this explosion been that observers could see its glow in broad daylight. In a few months, the Australian had identified the source of the explosion as the Crab Nebula in the constellation Taurus. This turbulent gaseous nebula is what remains of the nova observed by the Chinese, and is the result of highly agitated masses of gas interacting with each other.

Since the relatively unsung efforts of Jansky and Reber, hundreds of radio telescopes have sprung up all over the world, and this new branch of astronomy, unknown before Jansky's pioneer work, has yielded much dramatic and astounding information about the universe in which we live.

THE RADIO SUN

It has been found, for example, that the sun emits noise bursts over a wide frequency range, extending from the upper portion of the shortwave region to frequencies in the gigaHertz range. A close correlation between noise bursts and sunspot activity has been observed, and solar scientists have found good agreement between the types of noise bursts on the sun, particularly in the meter wavelength range, and ionospheric radio disturbances that follow certain types of emissions by one to two days. Solar flares, which occur chiefly during years of maximum sunspot activity, were also found to be active sources of radio noise. During periods of maximum sunspot activity, noise bursts up to one million times as intense as those observed during periods of low sunspot activity have been recorded.

RADIO PLANETS

Radio noise bursts strong enough to be detected have been observed from two planets, Venus and Jupiter, thus far. The Venusian noise, on a frequency of approximately 10 GHz, appears to be thermal in origin, and has provided information from which the range of temperatures on the planet can be deduced.

Noise bursts from the planet Jupiter are associated with massive thunderstorms, which appear to be related to the mysterious red spot on the planet. Additional radiation from Jupiter has led to the conclusion that it has a strong magnetic field, as well as radiation belts similar to the Van Allen radiation belts that surround Earth.

THE 21-CENTIMETER HYDROGEN LINE

In 1944 the Dutch astronomer van de Hulst predicted the discovery of radiation from space in the region 1400–1427 MHz, or a wavelength of 21 centimeters. His reasoning was based on the fact that this spectral line is emitted by hydrogen in its lowest rest energy level. The first observation of the existence in space of neutral atomic hydrogen was made in 1951 by two Harvard astronomers. Before radio astronomy there was no way of determining whether this gas, the most common element in the universe, was present in outer space. Subsequent observations on the 21-cm wavelength have shown it to be present in great abundance. Furthermore, by monitoring the shift in wavelength, it can be determined whether the hydrogen is moving toward or away from the observer. So important is this range of frequencies that the band has been allocated exclusively to the radio astronomy services by international agreement, and no other use by any service within the range 1400–1427 MHz is permitted.

Observations in the 21-cm region have yielded valuable data about the shape of the galaxy in which we live. The Earth is located in one of the spiral arms of a flat galaxy some 80,000 to 100,000 light-years across. It is some 33,000 light-years from the center of the galaxy. (A light-year is the distance light travels in an Earth year. Light travels at the speed of approximately 186,000 miles—300,000 kilometers—per second. In a year, a beam of light traveling at this speed will have traversed a distance of some 5,800 billion miles, or 9,000 billion kilometers.) It has been learned that the spiral arms of our galaxy are composed primarily of atomic hydrogen, and that these spiral arms begin at a distance of about 10,000 light-years from the center of the galaxy and extend at least 40,000 light-years from the center. Monitoring in the 21-cm region also indicates some hydrogen in our galaxy to be in an expanding and highly agitated state, and that much of the radiation coming to us from the galaxy has originated in regions where there are very powerful magnetic fields.

The 21-cm spectral line of hydrogen is significant in another respect: Many astronomers believe that technically advanced civilizations living beyond our solar system would attempt to communicate with other solar systems or galaxies using this wavelength. Thus far, monitoring of the 21-cm line has failed to yield any evidence of life in outer space. Exciting as the concept of life beyond Earth is to contemplate, little terrestrial effort has been made thus far to launch an organized search either on 21 centimeters, or any other wavelength.

QUASARS AND PULSARS

The two major astronomical discoveries of the past generation were made by radio astronomers. The first of these, made in the early 1960s, concerned very distant starlike objects which were found to be intense emitters of radio waves. Optically, the first of these objects to be discovered was presumed to be a star located in our own galaxy. Classified as 3C48 (the forty-eighth object to be classified in the Third Cambridge Catalog of Radio Sources), closer examination of the spectrum of 3C48 disclosed that it had a "red shift" corresponding to a velocity of about 30 percent of the speed of light, and an apparent distance of some 4 billion light-years. Other such objects have been located since, some of which appear to be at the outer limits of the observable universe, some 10 billion light-years away, and traveling at the apparent velocity of some 80 percent of the speed of light. Quasars, for Quasi-Stellar-Radio-Sources, have been one of the exciting astronomical discoveries of the century.

The second such discovery was made in 1967. Dr. A. Hewish and his associates at the Cambridge observatory discovered a source of pulsating radio waves, which ultimately was associated with a faint star. Subsequently, other such pulsating sources of radio energy were discovered. Having a period that varies from 1/30th to 2 seconds, these remarkable objects were found to have optical pulses in phase with the radio pulsations. The discovery of the *pulsars* caused great excitement among astronomers. No object that had ever been observed in the heavens behaved as the pulsars did, and it was at first felt by some that these regular beaconlike signals were being generated by intelligent beings. It was soon realized, however, that the power required to originate such pulses, of the order of 10^{26} watts, was far beyond the capability of any imaginable intelligence, and astronomers began to look elsewhere for an explanation.

It is now generally felt that pulsars are rapidly rotating neutron stars—the end product of stars that have consumed all their nuclear fuel and collapsed. Such stars are extremely small and dense. A neutron star would have a diameter of some 8 to 12 miles compared to our sun, whose diameter is some 800,000 miles. Neutron stars are extremely dense, of the order of 100 million tons per cubic centimeter, compared to about one ounce for the same amount of matter from our sun.

The emission spectrum of the average pulsar ranges between 40 and 2,000 MHz, and knowledge of this range has enabled astronomers to make a rough calculation of the distance to pulsars, all of which are presumed to be in our own galaxy. By assuming an electron density value for free space, and knowing that electrons in free space interact with electromagnetic radiation and slow it down, and that this slowing is more pronounced as the frequency increases, calculations can be made comparing the delay time between successive pulses at the high and low end of the radio spectrum emitted by each pulsar.

A second method of determining distances to pulsars was to monitor the pulsar radiation at frequencies near those that are known to be absorbed by hydrogen atoms in space. If there is hydrogen between the pulsar and the radio telescope monitoring it, some of the frequencies being received will be absorbed by the hydrogen. If no absorption takes place, it can be assumed that the pulsar lies within the spiral arms of the Milky Way. If energy absorption is observed, the pulsar lies either in or beyond one of the spiral arms of our galaxy. By using this method, the distances to pulsars have been determined to be between several hundred and several thousand light-years away.

RADIO ASTRONOMY AND COSMOLOGY

One theory dealing with the beginnings of our universe holds that some 12 billion years ago a cataclysmic explosion, or "big bang," started the universe on its present course of expansion. It has been postulated that the radiation from this cosmic blast would still exist in residual form in the microwave portion of the spectrum, in the form of unexplained noise. Such radiation has actually been observed on microwave frequencies from about 1 to 20 GHz. It is some one hundred times as strong as radiation in this region from known sources, and early indications have been that it is isotropic (coming uniformly from all directions) and gives strong support to the big-bang theory of the universe. If the omnidirectional background microwave noise radiation does indeed prove to be the residue of the big bang, a momentous contribution will have been made by radio astronomy to our understanding of the universe, and of how it began.

Fig. 83. National Radio Astronomy Observatory, Green Bank, West Virginia. At left is the 300-foot radio telescope, at right a 3-element interferometer. A 40-foot radio telescope is at left, rear. *Courtesy of the National Radio Astronomy Observatory.*

20

Lasers—Future Communicators

Almost from the beginning of radio communication scientists and technicians have tried to expand the spectrum of usable frequencies. In the earliest days, communications at longer and longer wavelengths were attempted. When in the early 1920s the amateurs revealed the promise of the high-frequency end of the then usable spectrum, more attention was paid to frequencies above 1.5 MHz. Then, in the mid-1920s, the dam broke, and the technical world found itself in possession of a broad band, ranging from 2 to 50 MHz, in which worldwide communication could be carried on at what was then considered fantastically low power. By the mid-1930s, use of the "shortwave band" had been reduced to practice, and the bulk of the world's wireless communications were being carried on in that region.

In the 1940s, ultra-high-frequencies (UHF) ranging up to 1,000 MHz (1 GHz) were brought under control, and microwave frequencies from 1 to 10 GHz began to be used.

The urge to increase the number of usable frequencies had another push when Claude Shannon published, with Warren Weaver, *The Mathematical Theory of Communication.* Practical men had realized, in a general way, that the more information was carried on a channel, the wider the bandwidth that was necessary. Shannon's work clarified and formalized that general impression, thereby introducing the new field of "information theory." It became apparent that with the urgent need for systems that carried a great deal of information, and therefore would require much greater bandwidth, it would be necessary to use higher and higher frequencies. (The amount of increase as the frequency is pushed up is not always realized. But if we consider the present available radio spectrum to be 300,000 MHz, from the lowest usable frequencies to above commonly used ones, and that increasing the upper limit from 300,000 to 600,000 MHz would double the number of usable frequencies, the importance of the higher frequencies becomes apparent.)

In the 1950s and 1960s it became possible to generate and detect signals of 100,000 MHz and higher (wavelengths of a few millimeters) but the small size of the circuit components restricted the power that could be generated and transmitted, and the equipment became difficult to manufacture. The miniature resonant cavities and tiny waveguides that were required showed that any higher frequencies would be a virtual impossibility. The limit had been reached.

But while most research scientists were concentrating on pushing the frequency frontier higher by conventional electronic techniques, a Hughes Aircraft Corporation scientist, Theodore Maiman, succeeded in producing a beam of pure red light, at a single frequency. This beam of light was to start another new era in electromagnetic communications. For this was no ordinary beam of light. It was *coherent*.

To know what *coherent* means, it is necessary first to consider ordinary, or *incoherent*, light. The light from any ordinary source comes out in a jumble of random waves independent of and uncoordinated with each other. They interfere, reinforcing and canceling each other in a random fashion. While it is obviously possible to signal with such a beam (the heliograph and signal lights are the oldest apparatus for communication with electromagnetic waves), the beam is broad (as compared to coherent light) and unmanageable. Maiman's *laser* (Light Amplification by Stimulated Emission of Radiation) compares with conventional means of generating light much as a modern highly selective continuous-wave generator compares with the original spark transmitter.

Laser action was originally produced by Maiman in a small rod of artificial ruby ("doped" like a transistor) but can also be obtained in a gas-discharge tube that looks much like a fluorescent lamp. Some of the atoms in the rod are shocked or "excited" by injecting a photon (small bit or packet of light) into them. With this extra bit of energy the atom is uncomfortable, and usually emits it almost instantly. The photon thus released travels along the rod. Every time it strikes another "excited" atom, it releases another photon, and they travel together down the tube. The ends of the tube are silvered (one end completely, the other enough to reflect about 85 percent of the light that strikes it). When the photons, traveling in step, reach the end of the tube, they are reflected and go back along the way they came, picking up more photons. At one of the ends, some of the photons leave the tube through the partly silvered mirror, forming the laser beam. (Those photons that started at a wide angle from the axis of the tube escape through the side and are lost.)

The coherent beam has enormously valuable properties that make it fantastically superior to ordinary light for communications purposes. First, it is single-frequency, having the advantage that CW has over spark, and is more easily and effectively modulated than ordinary light. The beam is more intense than any previously produced. The power of the ruby laser (in short bursts) reaches 10,000 watts for a beam measuring less than a square centimeter in cross section. It spreads out much less rapidly than a beam of noncoherent light. No matter how perfectly ordinary white light is focused into a beam by lenses or mirrors, it spreads out widely over the

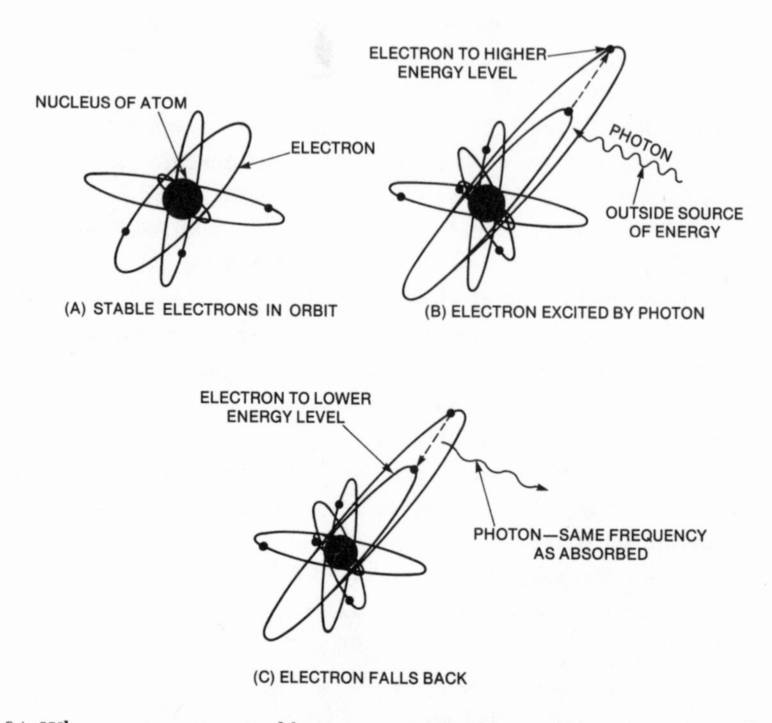

Fig. 84. When an atom is excited by an external force, one of its atoms may move to a higher energy (more distant) orbit. If the excitation is due to absorbing a photon of light, it drops to its nonexcited (ground) state by emitting the photon.

first few hundred yards or few miles. With a simple focusing lens, it is possible to project on the moon a laser light spot only two miles across.

Until the laser was developed, it had not been possible to generate frequencies above about 300 GHz. Then, in one step, more spectrum space became available than in all other bands combined. Frequencies in the visible and infrared region of the spectrum range from 430 to 750 *million* megaHertz, more than a thousand times the present available radio spectrum.

Laser devices are able to produce coherent radiation in many parts of the visible spectrum, as well as in the infrared portion. (Early lasers radiated only in the infrared.) The number of frequencies at which lasers have produced coherent radiation has been increasing steadily with continuing research. This trend is expected to continue.

Since the first working laser was announced, hundreds of laboratories in this country alone have joined in laser research. Television pictures have been transmitted over a beam of coherent light, and a number of laser systems have been used successfully in short-range experimental communication systems. Laser space-communications systems have been under test for several years.

Laser communication is, of course, strictly line-of-sight. This poses no problems for space communication, but can be inconvenient in many terrestrial applications. The probable solution to the problem is the *optical waveguide*, a thin fiber of special glass, many times as transparent as that used in window panes or telescope lenses. Because the surface of the glass (the boundary between it and the air) is an excellent mirror (like the surface of the water to a bather looking up at it at an angle from below), the guides can be curved—the light in them does not have to travel in straight lines, but is internally reflected from the inside surface as the guide bends.

Optical waveguides have been used in experimental cable-television systems in Japan and England, and parts of at least one telephone central station in the United States have been converted experimentally to laser transmission and optical waveguides. One of the chief obstacles to the greater use of fiber guides today is economic—the cost of the optical fibers is still high.

THE GAS LASER

In February 1961, scientists at Bell Laboratories announced the first continuous operation of a gas laser. Although structurally much different

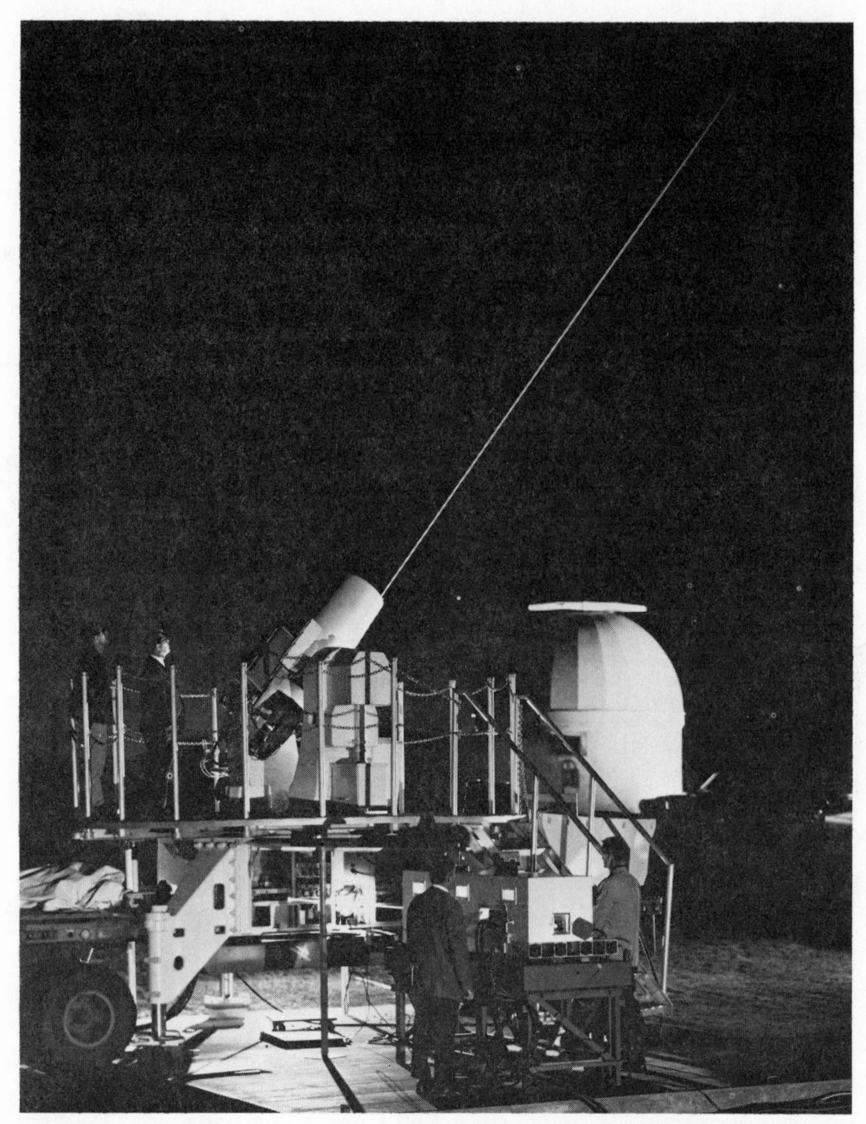

Fig. 85. Engineers of the Goddard Space Flight Center, Greenbelt, Maryland, operate a gas laser in experiments to communicate with a satellite in orbit. *Courtesy of NASA.*

from the solid-state laser, the basic principles were the same. The first gas laser was a quartz tube about 80 centimeters long and 1.5 centimeters in diameter. It contained a mixture of 90 percent neon and 10 percent helium gas, at a pressure of 1 to 2 millimeters of mercury. It produced coherent infrared emissions at five frequencies, the strongest at 11,530 angstrom units.

The earliest gas lasers followed the ruby type in many details, including silvered mirrors at the ends. These were soon replaced with clear glass windows and the mirrors placed outside the tube, where they could be adjusted more easily.

A radio-frequency generator excites the atoms in the gas laser. The electrical discharge excites the helium atoms, raising them to a higher energy level, as indicated in Figure 85, though the action is somewhat more complex than in the ruby laser. The helium atoms collide with the neon atoms in the tube, which are the ones that actually emit the photons that form the laser beam.

The radio-frequency generator makes it possible to operate the gas laser continuously. In the ruby laser, the exciting light is produced by a large gas-discharge flash tube, which produces so much heat that it can be operated only in short pulses.

THE INJECTION LASER

In November 1962, an entirely new concept in the production of coherent radiation was announced almost simultaneously by IBM, GE, and MIT. A new device, called an injection laser, employed a semiconductor diode driven directly by an electric current.

The injection laser consists of gallium arsenide doped with tellurium and zinc, to produce n- and p-type regions. These regions are joined as shown in Figure 86.

When current is applied, electrons move across the junction into holes. This is called *recombination*, and a photon is emitted on each recombination.

If the forward bias applied to the semiconductor is great enough, a large number of electrons and holes concentrate in a very narrow area, the *active region*, about 0.0001 of an inch wide on the p-side of the junction.

In the active region large numbers of photons are emitted. These, in turn, stimulate the emission of more photons by accelerating the recombination of injected electrons with holes. Each time a photon stimulates the emission of a second photon, the emission occurs in phase with the first, and in the same direction.

Since the thickness of the active region is so small, emitted radiation propagates most strongly in the plane of the junction. Waves traveling along the long axis remain in the junction region longer than any others. The rear face can be polished, as it is with the ruby laser, to obtain unidirectional action. The side faces of the laser are usually sawed or etched to permit radiation to pass in this direction with a minimum of internal reflection.

Early injection laser models operated at extremely high-current densities, of the order of 10,000 amperes per square centimeter. These models produced their light in pulses and could not operate continuously. Subsequently, CW injection lasers were developed to operate at much lower current densities, of the order of 100 A/cm^2.

Researchers have found other semiconductor materials that will *lase*. These include indium phosphide, indium arsenide, indium antimonide, and a gallium arsenide–gallium phosphide compound. Frequency ranges of early injection lasers extended from 7,000 angstrom units for the gallium arsenide phosphide compound to 52,000 angstrom units for indium antimonide. The frequency ranges produced by injection lasers run from 60 to 430 million megaHertz. These frequency ranges are in the infrared.

The most significant advances in injection lasers have come in the field of communications. Laser light is well suited to communication because it is emitted in nearly parallel beams, allowing maximum transfer of energy.

Thus far, solid-state lasers, such as the ruby, as well as the gas, have presented difficulties in modulating the light. An injection laser is relatively free of this difficulty, since the intensity of the light output is a function of the current in the laser once the semiconductor has begun to lase. Increasing the current increases the light output. Since the injection laser

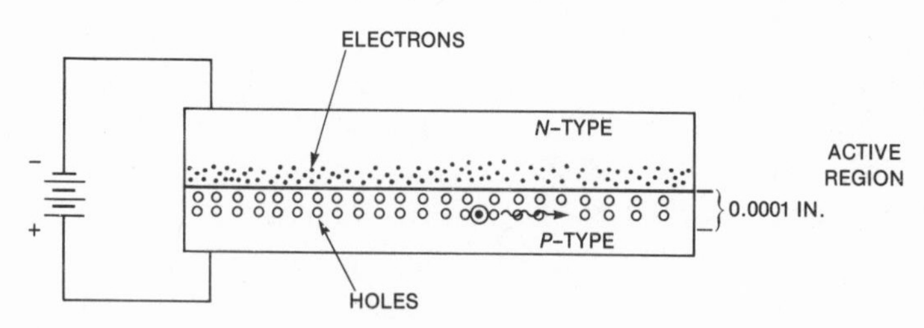

Fig. 86. Action of a solid-state injection laser.

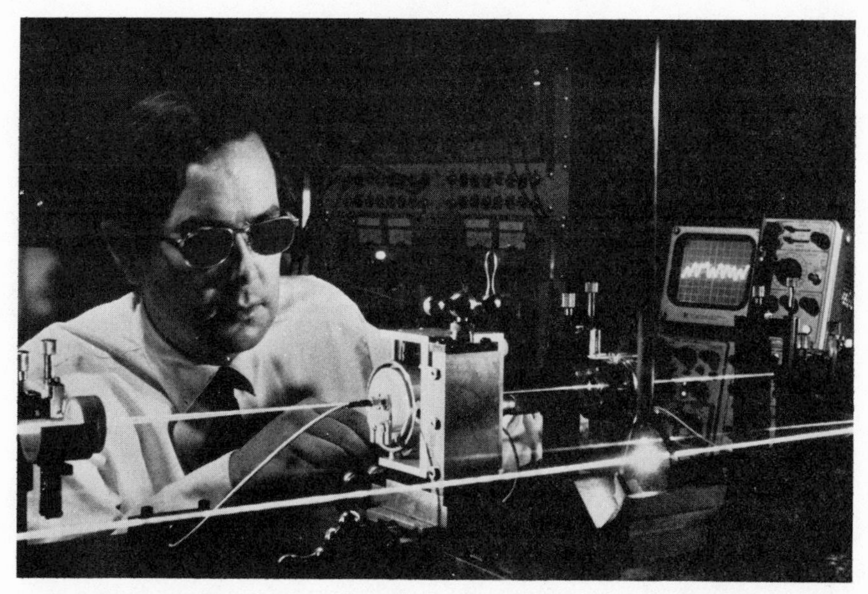

Fig. 87. Gerard White of Bell Telephone Laboratories, with an optical modulator, which can modulate a laser beam with voice, data, or other signals. *Courtesy of Bell Telephone Laboratories.*

can respond to driving current changes in a nanosecond (a billionth of a second), injection laser light can transmit up to 1 billion units of information in 1 second.

The apparatus consists of two basic components: the laser transmitter and its modulation circuitry, and the receiver, a phototube and demodulation circuitry. Because it is small, light, and more efficient than optically pumped solid-state and gas lasers, the injection laser is ideally suited for space-communications systems, and fits easily into an Earth satellite.

The small size of the injection laser, although advantageous, also presents some drawbacks. The region in which lasing action occurs is very small, because electrons, once they have crossed the junction, tend to drop into holes immediately. Since they move 0.0001 inch or less before recombination occurs, the power of the injection laser is limited. A second limitation is beam width. Although the injection laser produces highly directional beams, they still diverge significantly more than those produced by other lasers, particularly the gas type. Beam widths of the order of degrees are often produced by injection lasers, compared with a fraction of a degree for the gas laser.

THE FUTURE

Although lasers have been used to carry information from one point to another, costs and efficiency have so far precluded their use commercially. Before the vast bandwidth available in the visible portion of the spectrum can be used economically, efficient means of generating, amplifying, modulating, filtering, and detecting light will have to be developed. Thus far, the demands for additional bandwidth in the visible portion of the spectrum have not been great enough to warrant an all-out effort to develop systems which utilize the visible portion of the spectrum.

There seems little doubt that in time much of the world's communication requirements will be met by systems using coherent light to carry information. It is not clear what form those systems will take. The consensus among most scientists and engineers working in the field, however, is that the invention of the laser is one of the most important technological developments of the century, and will have a far-reaching and profound effect on communications of the future.

References

Angelakos, D.J., and Everhart, T.E. *Microwave Communications.* McGraw-Hill, New York, 1968.

Archer, G.L. *History of Radio to 1926.* The American Historical Society, Inc., New York, 1938.

Baker, W.J. *A History of the Marconi Company.* St. Martin's Press, New York, 1971.

Briggs, Asa. *The History of Broadcasting in the United Kingdom,* vol. I. Oxford University Press, London, 1961.

Chester, G., and Garrison, G.R. *Television and Radio.* Appleton-Century-Crofts, Inc., New York, 1956.

Cocking, W.T. "Milestones in Receiver Evolution." *Wireless World,* April 1971, pp. 160–63.

Comsat Report to the President and the Congress at the 10th Anniversary of the Communications Satellite Act of 1962.

Cook, J.S. "Deep Space Communications." *Bell Laboratories Record,* August 1970, pp. 213–18.

De Forest, Lee. *Father of Radio: The Autobiography of Lee de Forest.* Wilcox and Follett Co., Chicago, 1950.

DeSoto, Clinton B. *Two Hundred Meters and Down.* The American Radio Relay League, Inc., West Hartford, Conn., 1936.

Devereux, F.L. "Loud and Clear—Developments in Audio over 60 Years." *Wireless World,* April 1971, pp. 156–59.

Dictionary of Scientific Biography, vols. I–VI. Charles Scribner's Sons, New York.

Dunlap, Orrin E., Jr. *Radio's 100 Men of Science.* Harper and Brothers, New York, 1944.

Edelman, P.E. *Experimental Wireless Stations.* The Norman W. Henley Publishing Co., New York, 1920.

Electronic Design Magazine, vol. 20, no. 24. Nov. 23, 1972, devoted to the twentieth anniversary of the transistor.

Fifty Years of ARRL: A Reprint of Historical Articles from QST Magazine. The American Radio Relay League, Inc., Newington, Conn., 1965.

From Semaphore to Satellite. ITU Publication, Geneva, 1965.

Grambling, W.A., and Smith, R.C. "Laser Applications in Electronics." *Wireless World,* August 1969, pp. 367–71.

A Guide to Satellite Communication. Reports and Papers on Mass Communication. UNESCO Report no. 66, 1972.

Hamsher, D.H. *Communication System Engineering Handbook,* McGraw-Hill, New York, 1967.

Hannay, N.B. "The Solid State," *Annual Review of Physical Chemistry,* vol. 13. 1962, pp. 305–24.

History of Broadcasting and KDKA Radio. Undated and uncredited publication, Westinghouse Broadcasting Co.

Jacobs, George. "Guglielmo Marconi and the Sixtieth Anniversary of Transatlantic Wireless Communication." *CQ* Magazine, December 1961.

Jaffe, L. *Communications in Space.* Holt, Rinehart, and Winston, Inc., New York, 1963.

Kompfner, R. "Optical Communications." *Science,* vol. 150, no. 3693, October 8, 1965, pp. 149–55.

Laughter, V.H. *Operator's Wireless Telegraph and Telephone Handbook.* Frederick J. Drake and Co., Chicago, 1909.

Leinwoll, S. *Understanding Lasers and Masers.* John F. Rider, New York, 1965.

———. *Shortwave Propagation.* John F. Rider, New York, 1959.

———. *Space Communications.* John F. Rider, New York, 1964.

Maclaurin, W.R. *Invention and Innovation in the Radio Industry.* The Macmillan Co., New York, 1949.

Marconi, G. *Radio Telegraphy.* Paper presented before a joint meeting of the IRE and The American Institute of Electrical Engineers, June 20, 1922.

Miller, S.E. "Communication by Laser." *Scientific American,* vol. 214, no. 1, January 1966, pp. 19–27.

Proceedings of the IEEE: Special Issue on Radio and Radar Astronomy, vol. 61, no. 9, September 1973.

The Radio Amateur's Handbook. American Radio Relay League, Newington, Conn., 1972.

Radio Spectrum Utilization. A report of the Joint Technical Advisory Committee of the IEEE, New York, 1964.

Reference Data for Radio Engineers. 5th ed. Howard W. Sams, New York, 1970.

Satellite Communications Physics. Staff of the Bell Telephone Laboratories, 1963.

Satellites at Work: Space in the Seventies. NASA, 1970.

Schmeling, D. "Amateur Radio Frequency Allocations and Use." *QST* Magazine, April 1966, pp. 61–65.

Shawy, G.A. "From Morse to Satellites." *RCA Relay,* September/December 1967.

"A Short Wave Regenerative Receiver." Uncredited article, *QST* Magazine, December 1916, pp. 16–21.

Shunaman, Fred. "Lee de Forest, Father of Radio." *Radio-Electronics,* August 1973, pp. 52–55.

———. "The Transistor—25 Years Old." *Radio-Electronics,* December 1972, pp. 35–38.

Smith-Rose, R.L. "Fifty Years Research in Radio Wave Propagation." *Wireless World,* April 1961, pp. 203–7.

Stoffels, R.E. "Let's Talk Transistors." Three-part series, *QST* Magazine, November 1969–January 1970.

Terman, F.E. *Fundamentals of Radio.* McGraw-Hill, New York, 1938.

Van Allen, J.A. *Scientific Uses of Earth Satellites.* University of Michigan Press, 1956.

Verrill, A.H. *Harper's Wireless Book.* Harper and Brothers, New York, 1913.

Wellman, W.R. "Men of Radio." Five-part series, *CQ* Magazine, January, March, July, August 1952, January 1953.

Wymer, Norman. *Guglielmo Marconi.* The Marconi Company, Ltd., n.d.

Index